<u>Disclaimer</u>

The publisher of this book is by no way associated with the National Institute of Standards and Technology (NIST). The NIST did not publish this book. It was published by 50 page publications under the public domain license.

50 Page Publications.

Book Title: Development of a Method to Measure the Energy Consumption of Automatic Icemakers in Domestic Refrigerators with Single Speed Compressors

Book Author: David A. Yashar;

Book Abstract: This study examines the energy consumption of automatic ice makers installed in domestic refrigerators. This study builds upon the findings of a previous study and examines two refrigerator-freezers of different configurations, one French-door units with bottom freezers and one bottom mount unit that uses a twist tray mechanism to free frozen ice from the icemaker. Ice maker energy consumption is difficult to measure because they operate on a periodic cycle which is independent of the compressor cycle used to maintain the cold temperatures in the domestic refrigerator where it is installed; therefore methods proposed prior to this study have been subject to significant truncation error due to partial ice maker or compressor cycling. The purpose of this study is to define a method of measuring the energy consumption of automatic ice makers that will generate a repeatable and reproducible result. Several sets of test data from these units were analyzed and used to decipher the energy consumption of automatic ice makers. Through this effort, we developed a method of test to characterize ice maker energy consumption which circumvents the inherent problem with its measurement, truncation error due to incomplete cycling. The truncation error is avoided by measuring specific parameters with different sections of data from the same data set. This method was found to rapidly approach steady state values for the ice maker energy consumption. We then analyzed data sets from a prior study and found similar results for the stability of the ice making energy consumption; that continuous data over only 6 or 7 ice making cycles are typically sufficient to accurately characterize the energy consumption.

Citation: NIST TN - 1759

Keywords: energy consumption; ice maker; refrigerator

NIST Technical Note 1759

Development of a Method to Measure the Energy Consumption of Automatic Icemakers in Domestic Refrigerators with Single Speed Compressors

David A. Yashar, PhD

http://dx.doi.org/10.6028/NIST.TN.1759

NIST Technical Note 1759

Development of a Method to Measure the Energy Consumption of Automatic Icemakers in Domestic Refrigerators with Single Speed Compressors

David A. Yashar, PhD
Energy and Environment Division
Engineering Laboratory

http://dx.doi.org/10.6028/NIST.TN.1759

September 2012

U.S. Department of Commerce
Rebecca Blank, Acting Secretary

National Institute of Standards and Technology
Patrick D. Gallagher, Under Secretary of Commerce for Standards and Technology and Director

Certain commercial entities, equipment, or materials may be identified in this document in order to describe an experimental procedure or concept adequately. Such identification is not intended to imply recommendation or endorsement by the National Institute of Standards and Technology, nor is it intended to imply that the entities, materials, or equipment are necessarily the best available for the purpose.

National Institute of Standards and Technology Technical Note 1759
Natl. Inst. Stand. Technol. Tech. Note 1759, 51 pages (September 2012)
http://dx.doi.org/10.6028/NIST.TN.1759
CODEN: NTNOEF

DEVELOPMENT OF A METHOD TO MEASURE THE ENERGY CONSUMPTION OF AUTOMATIC ICEMAKERS IN DOMESTIC REFRIGERATORS WITH SINGLE SPEED COMPRESSORS

David A. Yashar

National Institute of Standards and Technology

Gaithersburg, MD 20899

Abstract

This study examines the energy consumption of automatic icemakers installed in domestic refrigerators. This study builds upon the findings of a previous study and examines two additional refrigerator-freezers of different configurations. Both of the units used in the present study, one French door unit and one single door unit, have a bottom-mounted freezer with an icemaker installed in the freezer compartment. The icemaker in the single door unit does not employ electric resistance heaters to free frozen ice from the icemaker, instead it uses a twist tray which consumes very little power. The energy consumption of icemaker is difficult to measure because it operates on a periodic cycle which is independent of the compressor cycle used to maintain the cold temperatures in the domestic refrigerator; therefore methods proposed prior to this study have been subject to significant truncation error due to parameters characterized over incomplete icemaker or compressor cycles. The purpose of this study is to define a method of measuring the energy consumption of automatic icemakers that will generate a repeatable and reproducible result.

Several sets of test data were analyzed and used to decipher the energy consumption of automatic icemakers. Through this effort, we developed a method of test to characterize icemaker energy consumption which circumvents the inherent problem with its measurement, truncation error due to incomplete cycling. The truncation error is avoided by measuring specific parameters with different sections of data from the same data set. This method was found to rapidly approach steady state values for the icemaker energy consumption. We then analyzed data sets from a prior study and found similar results for the stability of the ice making energy consumption; that continuous data over only 6 or 7 ice-making cycles are typically sufficient to accurately characterize the energy consumption.

After developing a method which eliminated the issues with truncation error, we examined the results and found that the energy consumption due to ice production is not a very strong function of the cabinet temperatures provided that (1) the test unit operates in such a manner that cabinet temperatures do not change in response to initiation of ice production, and (2) the unit does not operate near the limits of its cooling capacity during the test. This means that an accurate representation of the icemaker energy consumption can be acquired with a single data set if the temperature conditions during the ice making test are maintained near those measured during the baseline test.

Our measurements indicate that the ice-making energy consumption varied considerably between the units examined in this study. The most efficient product tested consumed approximately 0.177 kWh per kilogram of ice produced under the recommended test conditions resulting from this study, while the least efficient product consumed approximately 0.335 kWh per kilogram of ice. Variations in icemaker design and control algorithms played a large role in the ice-making energy consumption.

Keywords: energy consumption, icemaker, refrigerator

Acknowledgement

This work was sponsored by the United States Department of Energy, Office of Energy Efficiency and Renewable Energy, under the supervision of Lucas Adin. Dr. W. Vance Payne contributed to the test apparatus design, operation, and data acquisition. Mr. John Wamsley provided technician support throughout various phases of this project. Ms. Natascha Milesi-Ferretti of NIST and Dr. Detlef Westphalen of Navigant Consulting Inc. provided technical commentary on the draft of this report.

Table of Contents

List of Figures

List of Tables

1: Introduction

Cyclic type automatic icemakers are directly connected to a source of water and continuously produce batches of ice and store them in a low temperature bin. At the present, these devices are rendered inoperative during the regulated energy consumption test (10 Code of Federal Regulations Part 430, Subpart B, Appendix A1, 2010) therefore the energy consumed due to icemaker operation is not measured. This is because icemakers were not considered when the original basic energy consumption test (AHAM, 1979) was developed. The use of cyclic automatic icemakers do, however, have a significant impact on the product's energy use, which effectively goes undetected under the current regulatory test procedure. The objective of this study is to develop a robust, repeatable ice making energy consumption measurement method that could be applicable to cyclic automatic icemakers and would not substantially increase the test burden beyond that of the currently regulated energy consumption test.

Limited research exists in literature on energy consumption measurement of cyclic automatic icemakers. Meier and Martinez (1996) and Haider et al. (1996) studied the energy consumed by automatic icemakers. Although these studies were fairly preliminary, they provided a good starting point for a method of characterizing icemaker energy consumption. These studies have shown that the energy use associated with cyclic automatic icemakers can be quite substantial compared to the total energy used by a domestic refrigerating appliance. These studies also suggested that energy used to cool and freeze water into ice does not constitute the majority of their energy consumption.

More recently, NIST published a study (Yashar and Park, 2011) examining the energy consumption of four automatic icemakers installed in domestic refrigerators. Thier study provided a more thorough analysis of the units tested with the available methods, but was quite limited in the sampling of various technologies. The study provided a good starting point for the development of an icemaker energy test method, and included direction for future work needed to fully develop a test method. The current study builds upon the 2011 work by examining two other icemaker designs and providing more detailed analysis.

2: Test Setup and Data Acquisition

The test setup and data acquisition is described in Yashar and Park (2011); some of the major points are reiterated here. The setup was constructed in accordance with the Department of Energy's (DOE) test procedure outlined in 10 Code of Federal Regulations Part 430, Subpart B, Appendix A, 2010 with one slight modification, a continuous water supply was connected to supply water to the automatic icemaker. The test cells were located in an environmental chamber that is capable of providing controlled ambient temperature and humidity over long periods of time with little supervision, as necessitated by the lengthy test periods of domestic refrigerator energy consumption measurements.

Water supply lines were connected to each test unit. The temperature of the water supplied to each unit was not controlled, but the tube connecting each line to a refrigerator was sufficiently long to allow the water to equilibrate with the temperature in the chamber. This was verified by inserting two T-type thermocouples into each water line just upstream of the refrigerator connection.

All of the temperature and humidity data were gathered using a personal computer and a multiplexed data acquisition unit. The electrical energy input was monitored using a separate personal computer dedicated to two digital power meters, one connected to each test unit. All temperatures were sampled every 30 seconds, and the power was sampled every 2 seconds. Table 2.1 lists the measured quantities and the uncertainty associated with 95 % confidence. The equations used to calculate the measurement uncertainty are shown in the Appendix of Yashar and Park (2011).

Table 2.1: Measurement Uncertainty

Measured quantity	Measurement device	Uncertainty at 95 % confidence
Temperature	Thermocouples	$\pm$ 0.1 °C (0.2 °F)
Power	Watt-meter	$\pm$ 0.5 % of reading
Energy	Watt-meter	$\pm$ 0.5 % of reading
Mass	Digital scale	$\pm$ 5 g

Two domestic refrigerator-freezers were used for this study. The units were selected to provide information spanning a variety of designs and features that are expected to influence ice making energy consumption. The first unit in this study is a bottom-mount French door refrigerator-freezer with an icemaker located in the freezer compartment. The second unit is also a bottom-mount refrigerator-freezer with an automatic icemaker mounted in the freezer, but this unit employs a twist tray mechanism to remove frozen ice bits from the icemaker unlike most units which use electrical resistance heaters.

We examined the energy associated with ice making by comparing the average power drawn by the unit when it was and was not actively producing ice. The average power when the unit was not producing ice is a relatively straightforward parameter to measure. Determining the parameters while the unit was producing ice is more complicated because the icemaker typically operates in a cyclic fashion, but with a different period

2

from that of the compressor cycles. We therefore examined a few different options to characterize the icemaker energy consumption.

3: Experimental Results for French Door Unit

This unit was a 623 liter (22 cubic foot) French door energy star model without through the door (TTD) ice service, shown in Figure 3.1. This unit produces and stores ice in the freezer compartment, located in the drawer beneath the fresh food compartment, therefore the owner must pull out the freezer drawer in order to access the ice. This unit has a single speed compressor and it maintains the specified compartment temperatures by cycling the compressor on and off.

Figure 3.1- French Door Refrigerator Freezer without TTD Ice Service

3.1 Non-Ice Making Tests

The data presented in this section examines the steady state energy consumption of the test unit at various thermostat settings while the icemaker is inoperative. This data is necessary to assist in determining the influence of thermostatic settings on ice making energy. The steady state data was acquired in accordance with the procedure outlined in the 2014 U.S. Department of Energy test method. The steady state power consumption is characterized by taking the average power and compartment temperatures over a whole number of compressor cycles exceeding a time period of 3 hours. Multiple measurements are performed at different thermostat settings and the results are interpolated to determine the average power that would be used if the refrigerator was simultaneously maintaining prescribed target temperatures of 3.9 C in the fresh food compartment and -17.8 °C in the frozen food compartment.

3.1.1 Mid Setting Results

For the first set of tests we measured the energy consumption and compartment temperatures with the thermostats set to their median positions. The following results were obtained:

Steady state cyclic operation time: 13721 seconds = 03:48:41
Measured refrigerator compartment temperature: (2.3 ± 0.1) °C
Measured freezer compartment temperature: (-18.4 ± 0.1) °C
Energy expended during the test period: (216.7 ± 1.1) watt-hours

This yields an average steady state power of (56.7 ± 0.3) watts

3.1.2 Warm Setting Results

Since both compartment temperatures were colder than the target temperature during the first set of tests, we set the thermostats at to their warmest positions for the second measurement. The following results were obtained:

Steady state cyclic operation time: 15869 seconds = 04:24:29
Measured refrigerator compartment temperature: (7.6 ± 0.1) °C
Measured freezer compartment temperature: (-14.6 ± 0.1) °C
Energy expended during the test period: (201.9 ± 1.0) watt-hours

This yields an average steady state power of (45.8 ± 0.2) watts

3.1.3 Interpolating Steady State Results for Two Measurements

We interpolated the results of the first two measurements to determine the average power draw at the target temperatures. Using the refrigerator compartment temperature of 3.9 °C, the interpolation yields an average power of (53.4 ± 0.2) watts, while using the freezer compartment temperature of -17.8 °C yields an average power of (55.0 ± 0.3) watts. The higher of these two values is selected as the interpolation result in accordance with the existing test procedure.

3.1.4 Mixed Setting Results

We measured a third data point at a mixed thermostat setting in order to determine the average power at the exact target temperatures using the triangular interpolation method. We set the refrigerator compartment to the warmest setting and the freezer compartment to the coldest setting for this set of measurements. The following results were obtained:

Steady state cyclic operation time: 22506 seconds = 06:15:06
Measured refrigerator compartment temperature: (7.5 ± 0.1) °C
Measured freezer compartment temperature: (-22.2 ± 0.1) °C
Energy expended during the test period: (383.3 ± 1.9) watt-hours

This yields an average steady state power of (61.3 ± 0.3) watts

3.1.5 Interpolating Steady State Results for Three Measurements

We interpolated the results of all three measurements to determine the average power draw at the target temperatures. The triangular interpolation method (ASNZ, 2007) results in an average power draw of (54.5 ± 0.3) watts at the exact set of target temperatures.

3.2 Ice Making Tests

Once the baseline tests were completed, we performed a series of ice-making energy tests. In order to obtain the most representative data set, we measured the longest possible period of cyclic icemaker operation that we could obtain.

3.2.1 Mid Setting Results

For the first set of tests we measured the energy consumption and compartment temperatures with the thermostats set to their median positions. The data set used from this experiment is shown graphically in Figure 3.2 (power) and Figure 3.3 (temperatures). The data set that was selected for analysis encompasses the largest time period consisting of a whole number of compressor cycles where the unit operated in a steady periodic manner while producing ice. This set is indicated by the purple rectangle and includes 11 whole compressor cycles, during which time 16 batches of ice are harvested for a total of 1459 grams of ice. In this case, we have to work under the assumption that the same mass of ice is produced each time a batch of ice is harvested from the icemaker, 91.2 grams. A separate study would be needed to quantify the potential deviations from uniform batch mass that may affect the end result.

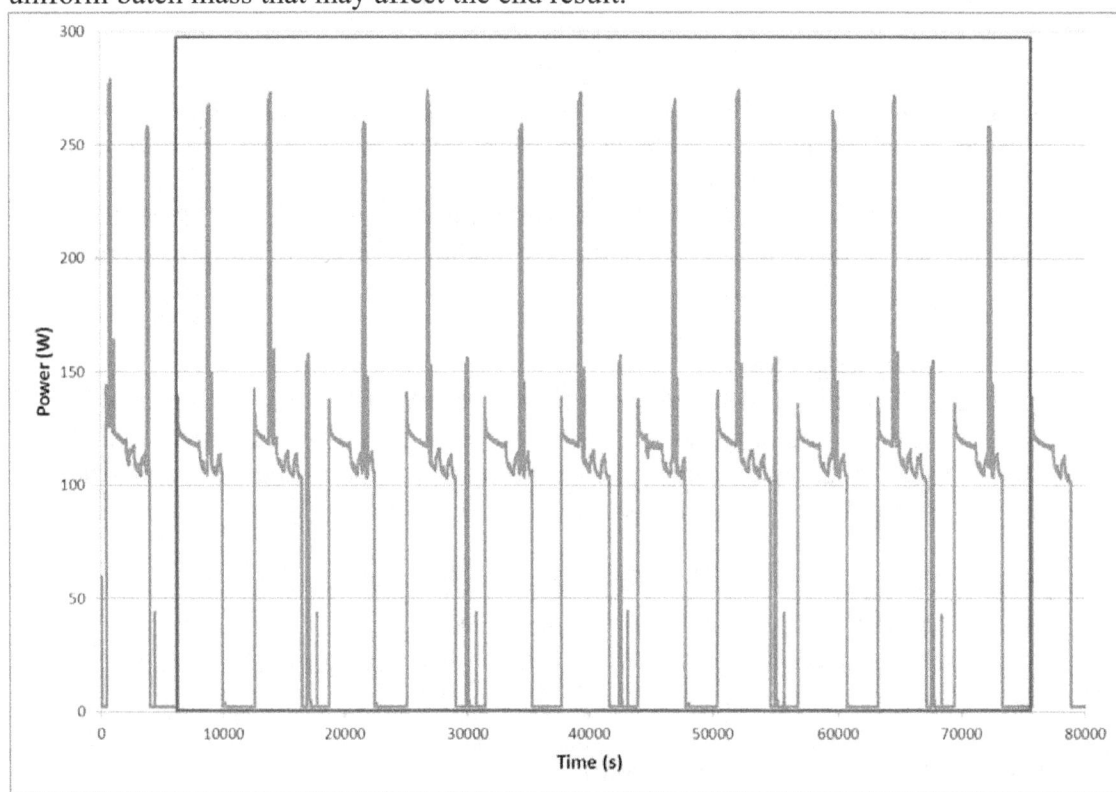

Figure 3.2 – Power at Mid Setting with Ice Production, French Door Unit

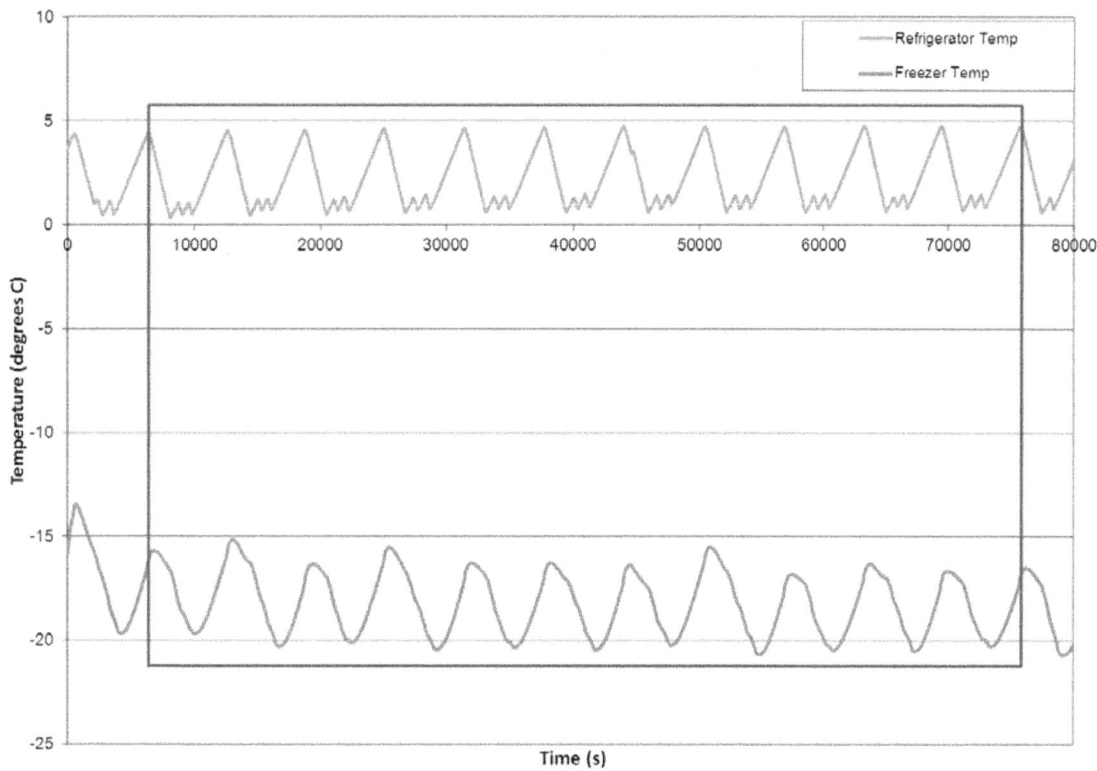

Figure 3.3 – Temperatures at Mid Setting with Ice Production, French Door Unit

Analysis yielded the following results:

Cyclic operation time: 69420 seconds = 19:17:00
Measured refrigerator compartment temperature: (2.2 ± 0.1) °C
Measured freezer compartment temperature: (-18.2 ± 0.1) °C
Energy expended during the test period: (1498.9 ± 7.5) watt-hours

This yields an average power of (77.7 ± 0.4) watts over the duration of this test period. One interesting aspect of the data pattern seen here is that while the ice making cycles and the compressor cycles are not synchronized, the compressor cycles have a much larger influence on the average power draw. This is because the magnitude and duration of the compressor power draw is much larger than that of the icemaker ejection heater. This means that a test period for the measurement of the average power should consist of whole compressor cycles. Figure 3.4 shows the cumulative average power draw versus time, relative to the beginning of the test period at the point when the compressor first switches on. This blue line indicates the values that would be obtained for the average power over the test period, if the end of the test period were selected from any point along the chart. The blue circles indicate the values that would be obtained if only whole compressor cycles were completed, i.e. the values at that would be obtained each time the compressor switches on.

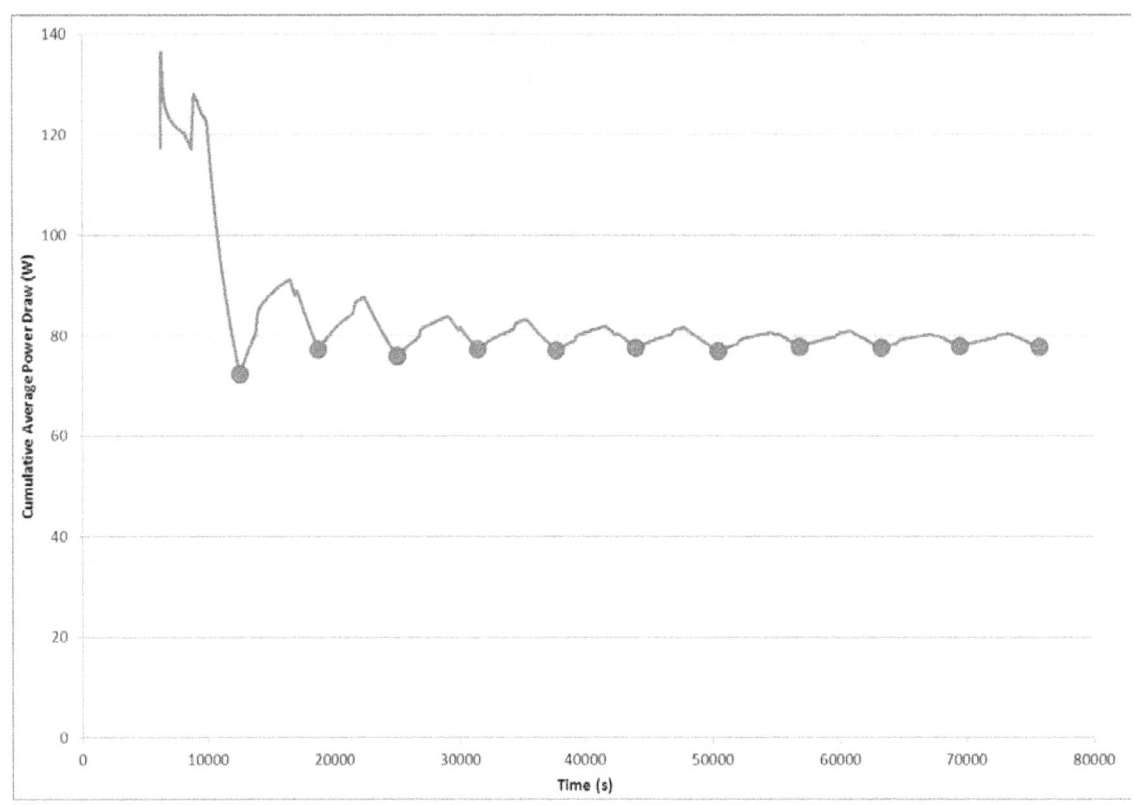

Figure 3.4 – Cumulative Average Power Draw at Mid Setting with Ice Production, French Door Unit

It is interesting to note that the values of the points indicated by the blue circles remain within 1 % of the value found using the full measurement period after the third compressor cycle. In other words, the average power draw could be determined within 1 % of the whole data set value by examining as little as four compressor cycles, in this case 25,129 seconds (6:58:49) of clock time.

If we consider the amount of ice produced during each harvest cycle as a basis for tracking the amount of ice at any given moment produced since the beginning of the test period, we can calculate the average energy consumed per mass of ice produced. Figure 3.5 shows the energy expended per mass of ice produced since the beginning of the test period. Again, the blue line represents the instantaneous values while the blue circles represent the points along the line that correspond to the values where the compressor starts, i.e. whole compressor cycle values.

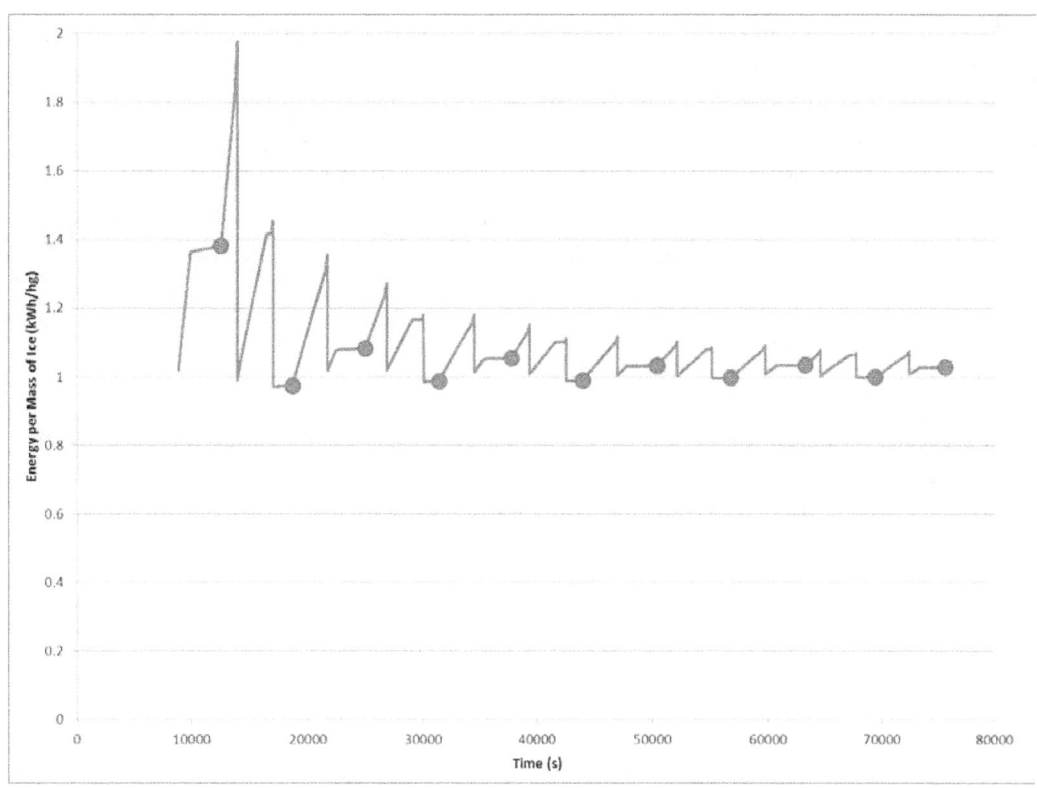

Figure 3.5 – Cumulative Average Energy per Mass of Ice Based on Compressor Starts, Mid Setting, French Door Unit

The figure shows this curve to progress in three distinct patterns: upward ramping, flat, and sharp drops. The upward ramping sections correspond to periods of compressor operation, when the unit is drawing considerable power. The flat sections correspond to periods of time when the compressor is not running, although some small amounts of energy may be consumed by other features. The sharp drops correspond to points when the icemaker harvests a batch of ice and the mass of ice produced realizes a sudden increase. Figure 3.5 shows that the compressor starts occur at fairly regularly time intervals, but the sharp drops to not, therefore the values of energy per mass of ice produced at whole compressor cycle intervals may occur at any random value along the blue line. As an example of the impact, an ice harvest event occurring immediately before or immediately after a compressor start would have a large influence on the measured value of the energy per mass of ice. Although the range of values bounded by the blue line becomes tighter with time, the values that may be obtained vary by +/- 3.5 % even after more than 19 hours of data collection.

The key to extracting useful information from this data is to determine the rate at which the unit produces ice. This is a rather difficult parameter to measure because the ice production rate is not regular. Figure 3.6 shows a segment of the data and highlights the variation between ice harvest cycles. This variation occurs because the icemaker harvests ice when it senses that the water in the molds has completely frozen, and the removal of heat from the water/ice in the molds occurs at a much higher rate during compressor-on periods than during compressor-off periods. The time required to produce a batch of ice will vary because the compressor-off periods slow down the process of freezing water.

The amount of energy per batch of ice, represented by the time integral of power between adjacent harvest events, appears to be a fairly stable value. This is a very important observation because the amount of energy per batch is directly coupled with the ice production rate and the average power required for producing ice while maintaining cabinet temperatures. Therefore, we can calculate a long term average of the ice production rate by dividing the average power by the energy consumption per mass of ice, which are quantities that are easier to measure.

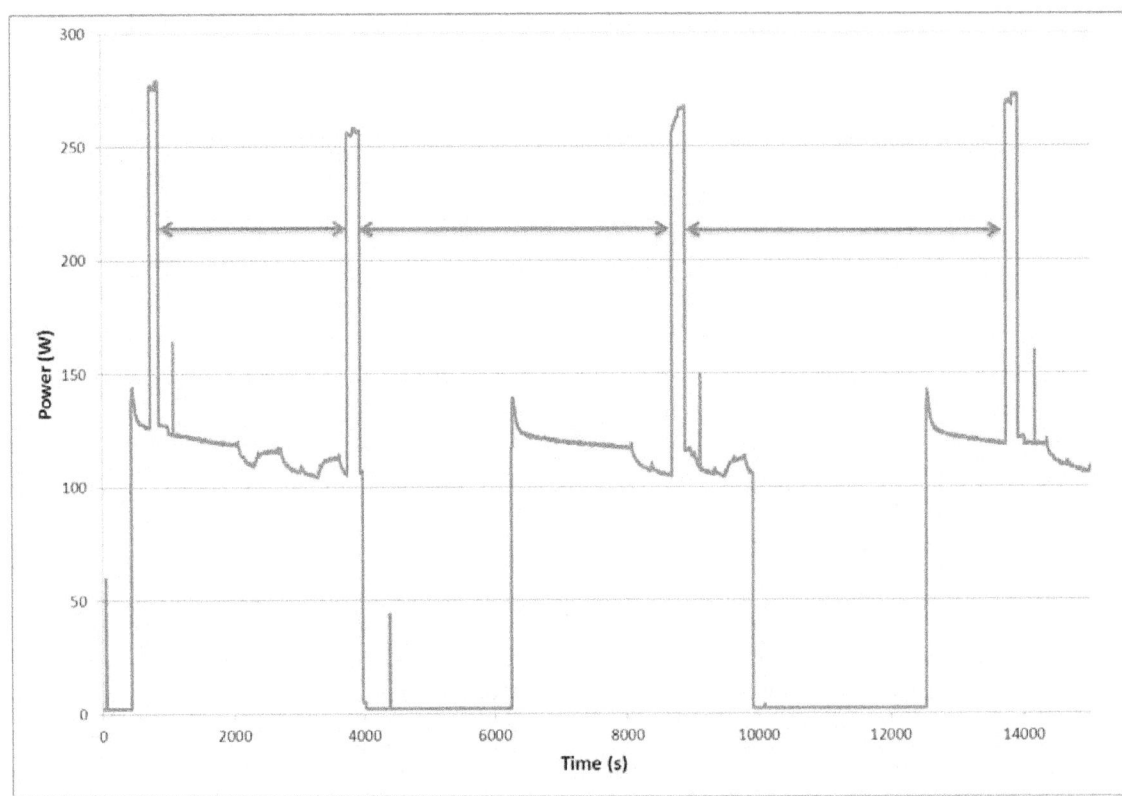

Figure 3.6 – Power at Mid Setting with Ice Production to Show Production Rate Variation

Figure 3.6 shows that the icemaker harvest heaters do not operate at regular intervals but the temporal spacing does seem to correlate with the compressor run time. This indicates that it would be better to measure the energy required to produce ice over whole ice making cycles. Figure 3.7 shows the same data as Figure 3.5, but this time the test period being analyzed starts and ends with the cycle of the ice harvest heaters. When analyzing the ice making energy data with this method, the values for energy per mass of ice produced at whole ice making cycles (blue circles) still have some variation, but not nearly to the same degree as those found when analyzing the data on a compressor cycle basis. Here, the variation is better than 1 % after only 8 ice harvest cycles, or 33,644 seconds (9:20:44) of clock time.

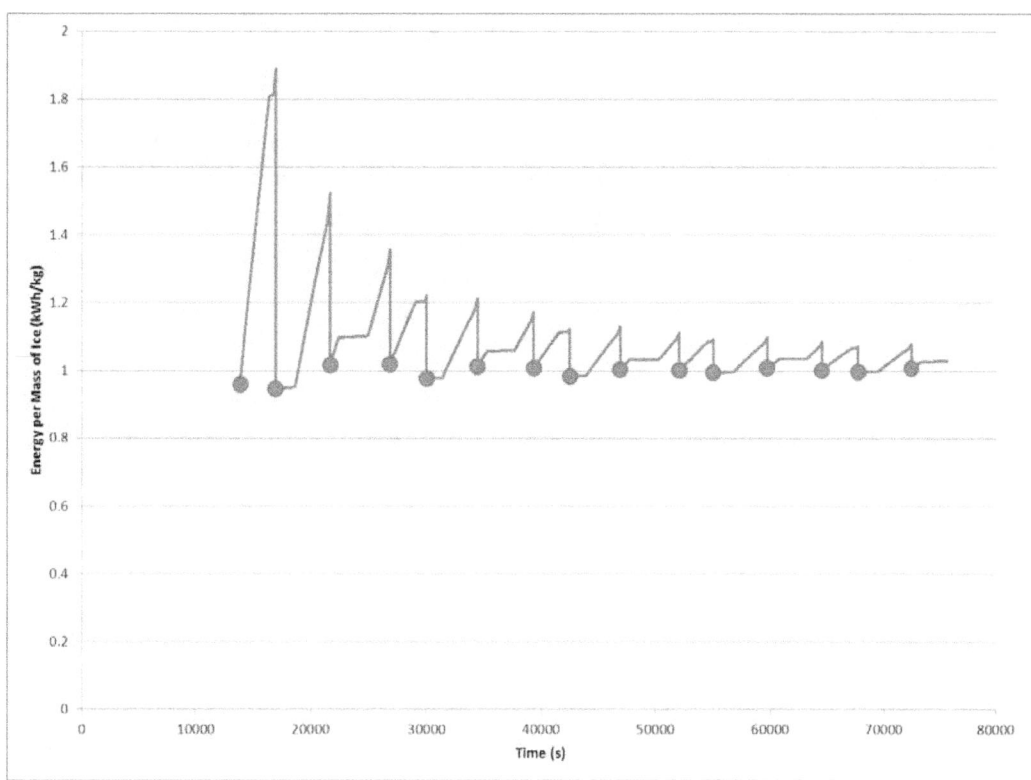

Figure 3.7 – Cumulative Average Energy per Mass of Ice Based on Ice Making Cycles, Mid Setting, French Door Unit

Using this method, we can determine that this refrigerator expends (1.007 ± 0.009) kWh of energy for each kilogram of ice produced while maintaining the internal cabinet temperatures. If we divide the average power of (77.7 ± 0.4) watts by this value, we obtain (0.0772 ± 0.0008) kg/h for the ice production rate. This is the value that we would expect the average to approach if the unit were able to operate, uninterrupted, for a very long time.

Now that we have the ice production rate and the average power for this data set, we can determine the ice making energy using the simplest method; the method in which only data from the median thermostat settings is used. It is important to note that this refrigerator did not undergo any significant change in compartment temperature due to the operation of the icemaker; therefore these data sets should make a good basis for comparison.

We first need to determine the average power increase due to the operation of the icemaker by subtracting the average power obtained from the test at the same thermostat settings.

(77.7 ± 0.4) W - (56.7 ± 0.3) W = (21.0 ± 0.5) W

Then we can determine the icemaker energy consumption by dividing this value by the ice production rate.

11

(0.0210 ± 0.0005) kW / (0.0772 ± 0.0008) kg/h = **(0.272 ± 0.007) kWh/kg**

3.2.2 Warm Setting Results

The next set of measurements was used to examine the effects of interpolating the results from multiple ice making energy consumption tests. We set the thermostats for each compartment to their warmest setting and collected data. Figure 3.8 and Figure 3.9 shows the power and temperatures measured during these tests, respectively.

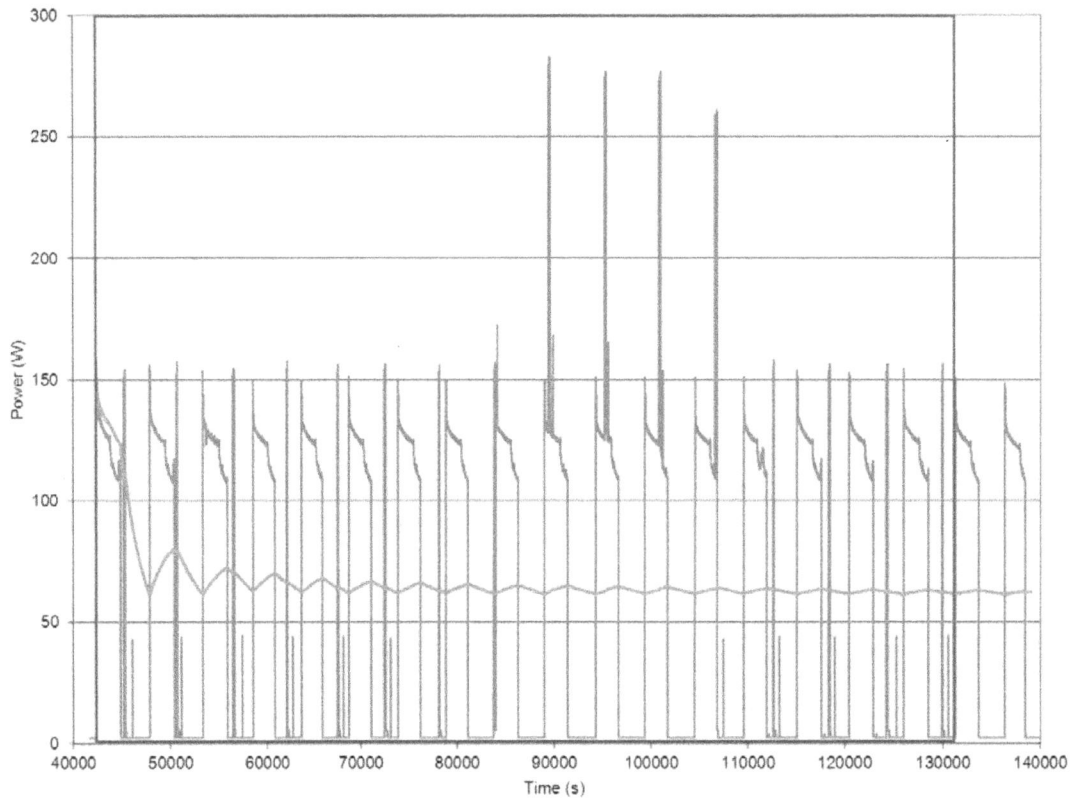

Figure 3.8 – Power at Warm Setting with Ice Production, French Door Unit

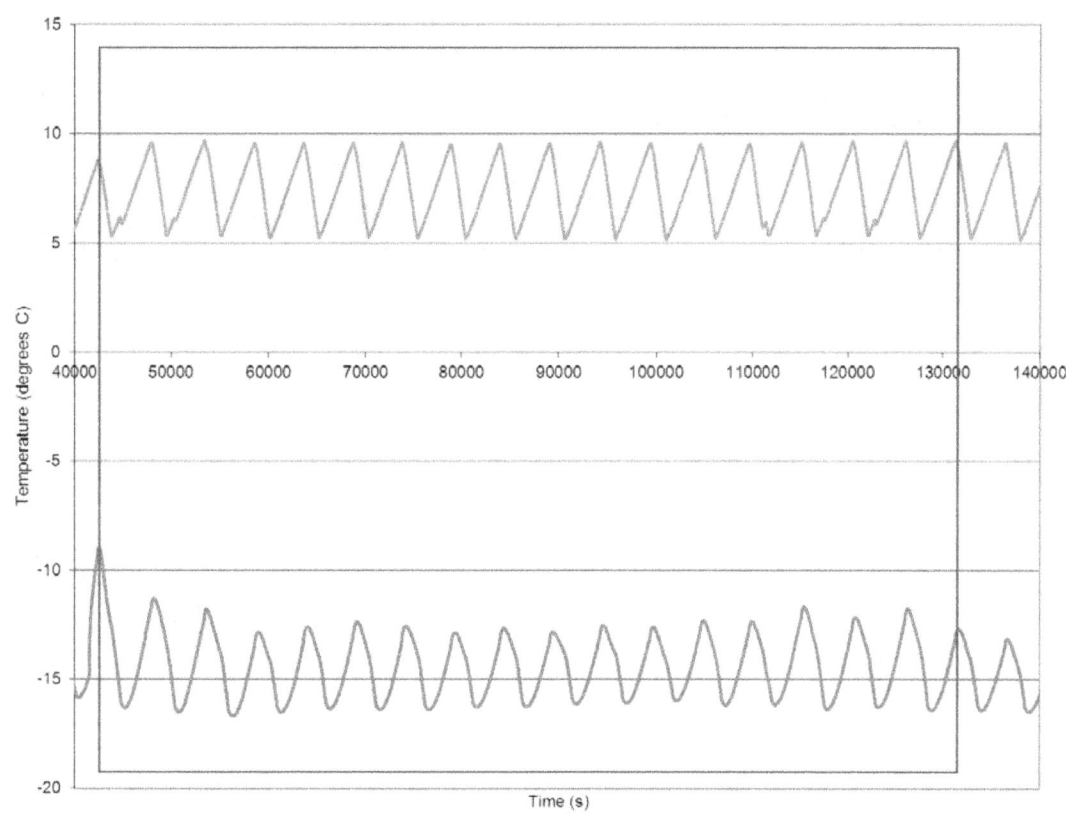

Figure 3.9 – Temperatures at Warm Setting with Ice Production, French Door Unit

Both figures show a purple rectangle around the test period, which is the largest obtainable data set during which the icemaker was actively producing ice. In total the data set encompasses 88,885 seconds (24:41:25) of clock time. This data set encompasses17 compressor cycles and 16 ice harvest cycles. Figure 3.8 shows both the instantaneous power (blue) and the cumulative average power (green) starting at the beginning of the test period. The following results were obtained from the data set:

Measured refrigerator compartment temperature: (7.4 ± 0.1) °C
Measured freezer compartment temperature: (-14.4 ± 0.1) °C
Energy expended during the test period: (1515.3 ± 7.6) watt-hours
Average power: (61.4 ± 0.3) W

Using the same analysis as performed on the previous data set, we can show the average power over a whole number of compressor cycles is stable to within 1 % of the full data set value after the fourth full compressor cycle, which means that this value could have been obtained with 26,253 seconds (07:19:12) of data.

Examination of the data shows that if the compressor cycles are used as the basis for measuring the energy per mass of ice produced, the resultant values which may be obtained can vary by +/- 5 % even after more than 24 hours of data collection. Figure 3.10 shows the energy per mass of ice produced using the ice making cycles as the basis for the test period. The total energy per mass of ice for the entire test period is

(1.076± 0.005) kWh of energy for each kilogram of ice produced. Here, the values vary by less than 1 % after only 11 ice harvest cycles, or 67,390 seconds (18:43:10) of clock time. This is not as fast as the previous data set, but it is much faster than the previously suggested 24 h time period (AHAM, 2009).

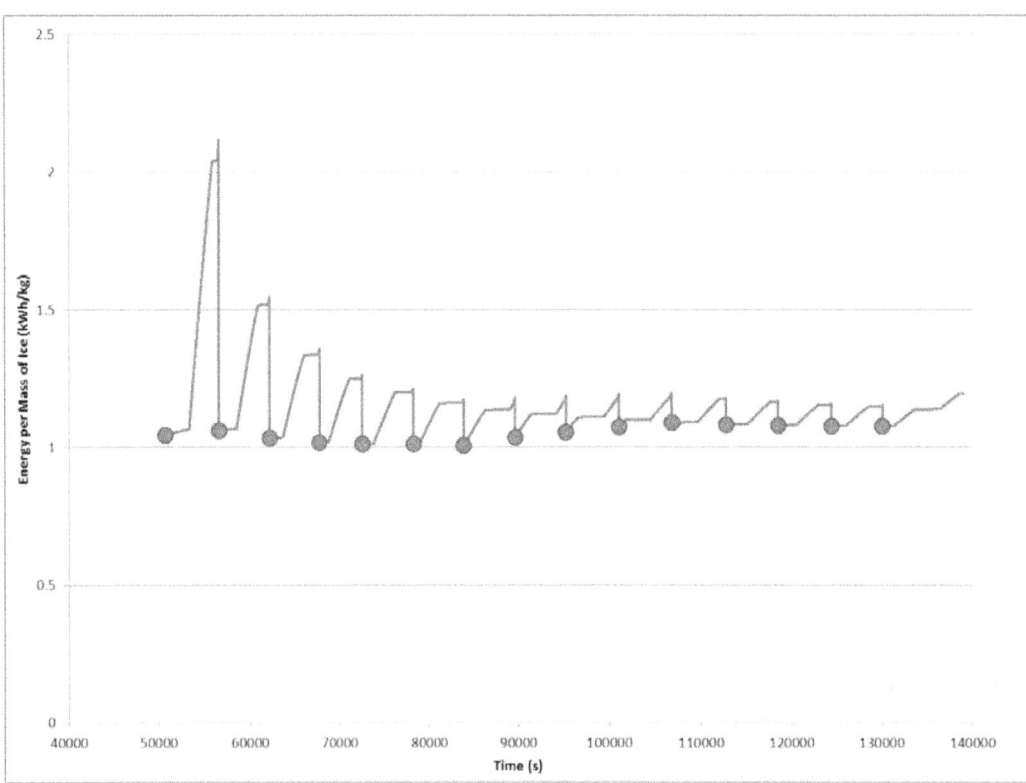

Figure 3.10 – Cumulative Average Energy per Mass of Ice Based on Ice Making Cycles, Warm Setting, French Door Unit

We can calculate the average ice production rate using the average power and the energy per mass of ice produced.

(0.0614 ± 0.0003) kW / (1.076 ± 0.005) kWh/kg = (0.0571 ± 0.0003) kg/h

Now that we have these parameters, it is interesting (although not necessary) to calculate the increase in power consumption due to ice making energy at the warmest thermostat settings; again it is noted that this unit did not exhibit a significant change in cabinet temperature in response to the operation of the icemaker. In this case, the incremental power increase is found by subtracting the average power with and without the icemaker operational.

(61.4 ± 0.3) W - (45.8 ± 0.2) W = (15.6 ± 0.4) W

Then we can determine the icemaker energy consumption by dividing this value by the ice production rate.

(0.0156 ± 0.0004) kW / (0.0571 ± 0.0003) kg/h = **(0.273 ± 0.007) kWh/kg**

14

In this case, the difference is minimal compared to the results of the median temperature setting test case.

3.2.3 Interpolating Ice Making Results for Two Measurements

Now that we have the ice production rate, the average power, and the compartment temperatures for data sets acquired at the median and warmest settings, we can determine the ice making energy using a more robust method. This method is similar to the current rating method for refrigerators in that it requires interpolation of data taken from multiple measurements in order to estimate the parameters at the target temperatures of 3.9 °C in the fresh food compartment and -17.8 °C in the frozen food compartment. The following results were obtained by interpolating the average power and ice production rate from the first two data sets to the frozen food compartment temperatures while exceeding the fresh food compartment temperature; the temperature conditions were therefore 2.7 °C and -17.8 °C.

Average power = (76.0 ± 0.4) W
Ice Production Rate = (0.0751 ± 0.0005) kg/h

Using these values and those obtained earlier from the non-ice making tests, we can calculate the incremental power increase attributed to ice making.

(76.0 ± 0.4) W - (55.0 ± 0.3) W = (21.0 ± 0.5) W

The ice making energy consumption is therefore:

(0.0210 ± 0.0005) kW / (0.0751 ± 0.0005) kg/h = **(0.280 ± 0.007) kWh/kg**

3.2.4 Mixed Setting Results

For the sake of completeness, we will also examine a data set collected at a mixed thermostat setting. We set the freezer temperature to the coldest setting and the refrigerator temperature to the warmest. Under these conditions, the compressor operated for longer periods of time than was seen in the other data sets in order to maintain the cold freezer temperature, therefore we were only able to obtain four full compressor cycles during which time the unit was actively producing ice. The blue line in Figure 3.11 shows the power signature of the unit under these test conditions. The green line in Figure 3.11 shows the running average power from the beginning of the test period, and the green circles show the full cycle values of the average running power.

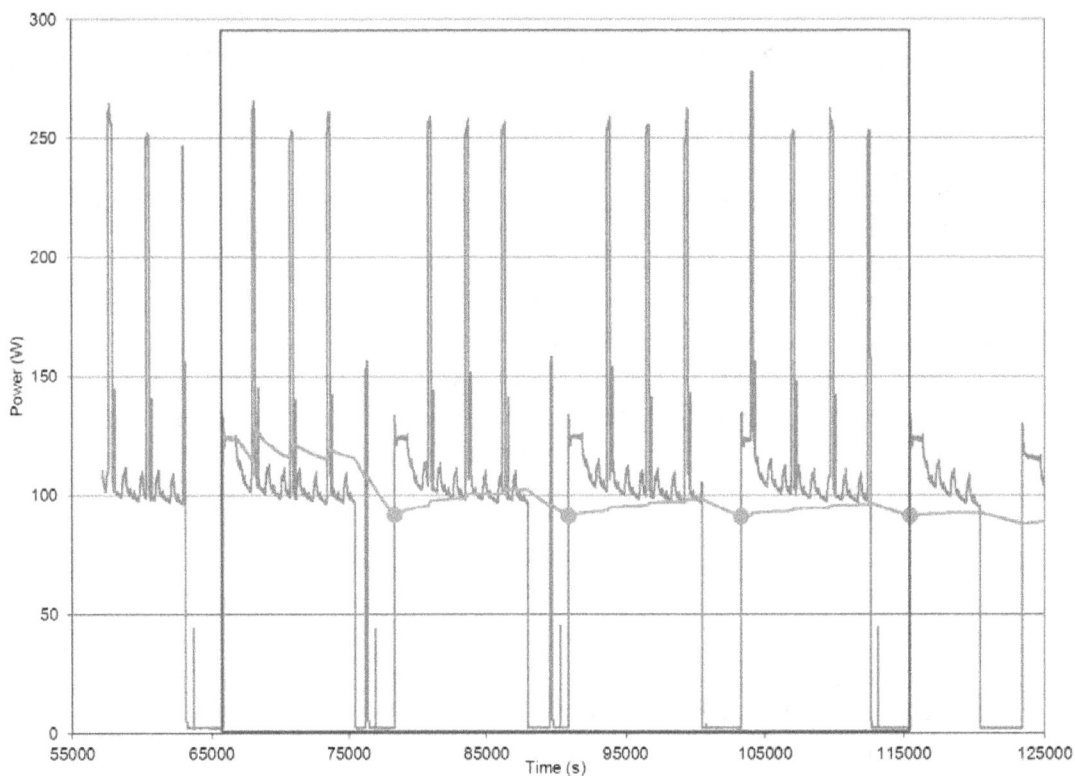

Figure 3.11 – Instantaneous and Cumulative Average Power Draw at Mixed Setting with Ice Production, French Door Unit

Although this data does not continue for enough cycles to satisfy the steady state criteria in the DOE test procedure, each of the four individual cycles in the data set has an average power that is within 1 % of the overall average power. Therefore, the average power could be determined from a test period that is as short as a single compressor cycle, or 12514 seconds (03:28:34). The necessary data from this set is:

Measured refrigerator compartment temperature: (7.7 ± 0.1) °C
Measured freezer compartment temperature: (-21.9 ± 0.1) °C
Energy expended during the test period: (1258.7 ± 6.3) watt-hours
Average power: (91.3 ± 0.5) W

Figure 3.12 shows the energy expended per total mass of ice produced during the test period. The blue line shows the cumulative average and the blue circles show the values obtained at whole icemaker cycle intervals.

16

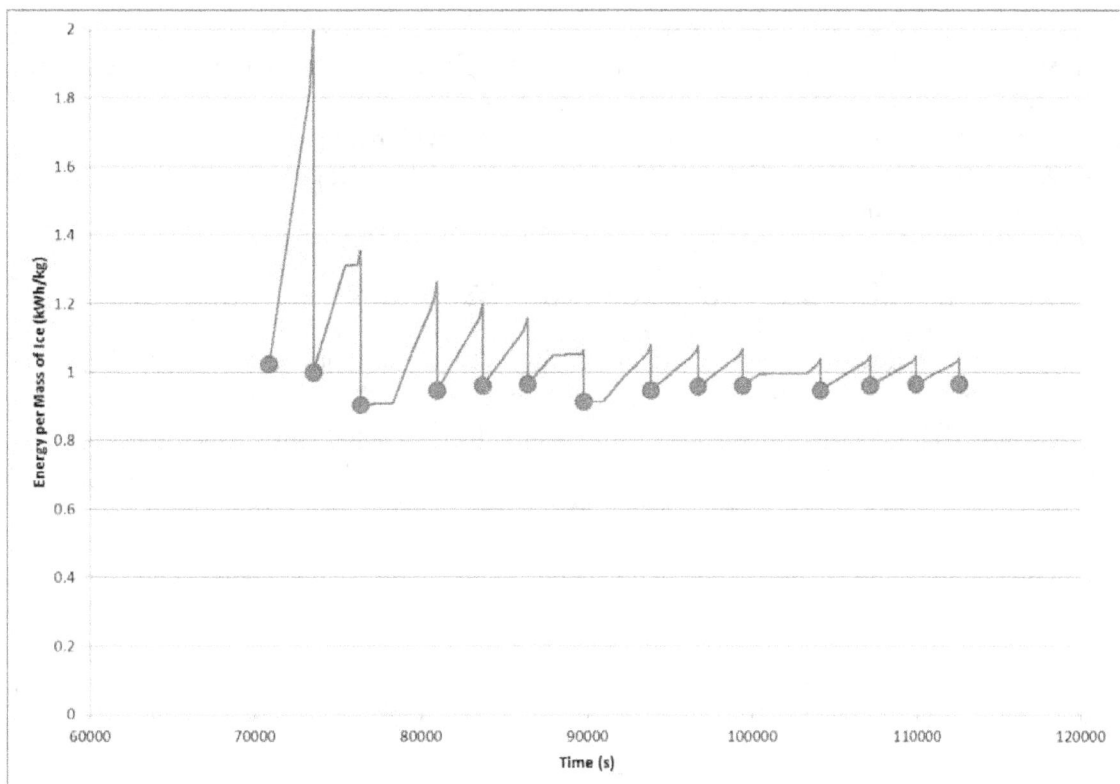

Figure 3.12 – Cumulative Average Energy per Mass of Ice Based on Ice Making Cycles, Mixed Setting, French Door Unit

Again, we can see some of the same trends for the energy per mass of ice produced. The running average of total energy per mass of ice produced was stable to within 1 % after 11 ice-making cycles. The final value for this parameter is (0.9668 ± 0.005) kWh of energy for each kilogram of ice produced.

Finally, the ice production rate can be calculated by dividing the average power by the energy to produce ice.

(0.0913 ± 0.0005) kW / (0.9668 ± 0.005) kWh/kg = (0.0944 ± 0.0007) kg/h

We can again examine the differential energy consumption due to ice making at another single set of thermostat settings with this data set since the compartment temperatures did not change significantly in response to the operation of the icemaker. In this case, the incremental power increase is found by subtracting the average power with and without the icemaker operational.

(91.3 ± 0.5) W - (61.3 ± 0.3) W = (30.0 ± 0.6) W

Then we can determine the icemaker energy consumption by dividing this value by the ice production rate.

(0.030 ± 0.0006) kW / (0.0944 ± 0.0007) kg/h = **(0.318 ± 0.007) kWh/kg**

17

In this case, the difference is larger than the results of the other single temperature setting test cases. However, it is important to note that this data set was collected while producing ice in a very cold frozen food compartment which caused the unit to operate with very little compressor off time. This indicates that the unit is operating in a more inefficient manner than it did under the other tests.

3.2.5 Interpolating Ice Making Results for Three Measurements
Lastly, we will assemble the three data sets using the triangular interpolation for both the power and ice production rates. As shown earlier, the average power that would be realized without ice production at a freezer temperature of -17.8 °C and a refrigerator temperature of 3.9 °C is (54.5 ± 0.3) W. We can apply the same calculation to the average power found during the ice making tests. In this case the calculated average power is (75.7 ± 0.4) W. Applying the same calculation to the ice production rate results in a value of (0.0748 ± 0.0007) kg/h.

Then, the differential power at the target temperatures is

(75.7 ± 0.4) W - (54.5 ± 0.3) W = (21.2 ± 0.5) W

And then the icemaker energy consumption is:

(0.0212 ± 0.0005) kW / (0.0748 ± 0.0007) kg/h = **(0.283 ± 0.007) kWh/kg**.

3.3 Summary
In this section we analyzed six data sets from one refrigerator-freezer operating under various conditions. Through these analyses, it was shown that it is necessary to use different test periods to extract all of the relevant parameters needed to quantify the ice making energy consumption. The refrigerator temperature, freezer temperature, and energy input must be measured over a whole number of compressor cycles; while the quantity of ice produced and the average energy per batch of ice must be measured over a different test period consisting of a whole number of ice making cycles. The ice production rate is not a stable parameter and must be estimated from the average power and the average energy per batch of ice.

We calculated the energy used to produce ice by 5 different means using the various methods of assembling this data. The table below summarizes the results.

Table 3.1: Summary of Test Results for French Door Unit

Method	Compartment Temperatures Fridge and Freezer (°C)	Ice Making Energy (kWh/kg)
Single Point	$T_{FR} = 2.2$; $T_{FZ} = -18.2$	0.272 ± 0.007
	$T_{FR} = 7.4$; $T_{FZ} = -14.4$	0.273 ± 0.007
	$T_{FR} = 7.7$; $T_{FZ} = -21.9$	0.318 ± 0.007
2 point interpolation	$T_{FR} < 3.9$; $T_{FZ} = -17.8$	0.280 ± 0.007
3 point interpolation	$T_{FR} = 3.9$; $T_{FZ} = -17.8$	0.283 ± 0.007

18

It is interesting to note that most of the values in the table are close to each other within the measurement uncertainty. There is one value that is a significant outlier, the single data set method using the mixed thermostatic setting. The operation of the unit under the mixed thermostat conditions was very different from the other data sets since the balance of cooling load between the refrigerator and freezer compartments was different.

4: Experimental Results for Bottom Mount Unit with Twist Tray

The second unit examined, shown in Figure 4.1, was quite different from most units that are commercially available in the United States at the time of this study. First of all, it is much smaller than most units with a capacity of 266 liters (9.4 cubic feet). It is configured with the refrigerator compartment on top and the freezer on the bottom; the icemaker is located inside a small drawer in the freezer. The unit has a multi speed compressor that operates at a few discreet settings and switches on and off to maintain the specified compartment temperatures. The unit also has two separate evaporators for the fresh food and frozen food compartments.

One particularly interesting aspect of this unit is that its icemaker operates in a very different manner to most other products. Most products free the frozen ice from the mold trays by using electric resistance heaters to melt the interface between the ice cubes and the molds. In this product, water is frozen into ice within a pliable mold tray. The removal mechanism is purely mechanical, a small amount of torque is applied to the tray which twists it and ejects the frozen ice cubes. This unit was specifically included in this study to examine whether units without ice mold ejection heaters would realize an inherent difficulty in measuring the icemaker energy consumption.

Figure 4.1 – Bottom Mount Test Unit with Twist Tray Icemaker

4.1 Non-Ice Making Tests

The data presented in this section examines the steady state energy consumption of the test unit at various thermostat settings. This data is necessary to assist in determining the influence of thermostatic settings on ice making energy. The steady state data was acquired in accordance with the procedure outlined in the 2014 U.S. Department of Energy test method.

4.1.1 Mid Setting Results

For the first set of tests we measured the energy consumption and compartment temperatures with the thermostats set to their median positions. The variations of the power draw due to the operation of the multi-speed compressor necessitated a somewhat longer steady state test period than most units according to the DOE test method. The following results were obtained:

Steady state cyclic operation time: 48495 seconds = 13:28:15
Measured refrigerator compartment temperature: (3.4 ± 0.1) °C
Measured freezer compartment temperature: (-18.5 ± 0.1) °C
Energy expended during the test period: (717.3 ± 3.6) watt-hours

This yields an average steady state power of (53.3 ± 0.3) watts

4.1.2 Warm Setting Results

Since both compartment temperatures were colder than the target temperature during the first set of tests, we set the thermostats at to their warmest positions for the second measurement. The following results were obtained:

Steady state cyclic operation time: 26136 seconds = 07:15:36
Measured refrigerator compartment temperature: (6.3 ± 0.1) °C
Measured freezer compartment temperature: (-15.8 ± 0.1) °C
Energy expended during the test period: (301.7 ± 1.5) watt-hours

This yields an average steady state power of (41.6 ± 0.2) watts

4.1.3 Interpolating Steady State Results for Two Measurements

We interpolated the results of the first two measurements to determine the average power draw at the target temperatures. Using the refrigerator compartment temperature of 3.9 °C, the interpolation yields an average power of (51.8 ± 0.3) watts, while using the freezer compartment temperature of -17.8 °C yields an average power of (51.0 ± 0.3) watts. The higher of these two values is selected as the interpolation result.

4.1.4 Mixed Setting Results

We measured a third data point at a mixed thermostat setting to in order to determine the average power at the exact target temperatures using the triangular interpolation method. We set the refrigerator compartment to the coldest setting and the freezer compartment to the median setting for this set of measurements. The following results were obtained:

Steady state cyclic operation time: 57498 seconds = 15:58:18
Measured refrigerator compartment temperature: (2.0 ± 0.1) °C
Measured freezer compartment temperature: (-17.1 ± 0.1) °C
Energy expended during the test period: (955.6 ± 4.8) watt-hours

This yields an average steady state power of (59.8 ± 0.3) watts

4.1.5 Interpolating Steady State Results for Three Measurements
We interpolated the results of all three measurements to determine the average power draw at the target temperatures. The triangular interpolation method results in an average power draw of (52.5 ± 0.5) watts.

4.2 Ice Making Tests
Once the baseline tests were completed, we performed a series of ice-making energy tests. Before examining the results, it is important to note that the data analysis for this unit was a little more complicated due to the absence of the ice mold ejection heater. In the more traditional units, the ice mold ejection heaters are typically operated for one to two minutes at a power ranging from 100 W to 200 W. Their operation is clearly visible upon examination of the power measurements and can be used to indicate the start and end points of an ice making cycle. This unit does not use such heaters, but instead uses a relatively low power device to twist the tray, and this makes it more difficult to identify the beginning and end points of an ice making cycle by monitoring the power to the unit.

Since the beginning and end points of the ice making cycles were not easily identifiable by examining the power measurement data, we used another measured parameter to assist. Throughout the course of this study, we measured the temperature of the water that was supplied to the unit. This measurement was taken at the inlet to the appliance. Figure 4.2 shows the inlet water temperature during the ice making data set acquired at the median temperature setting.

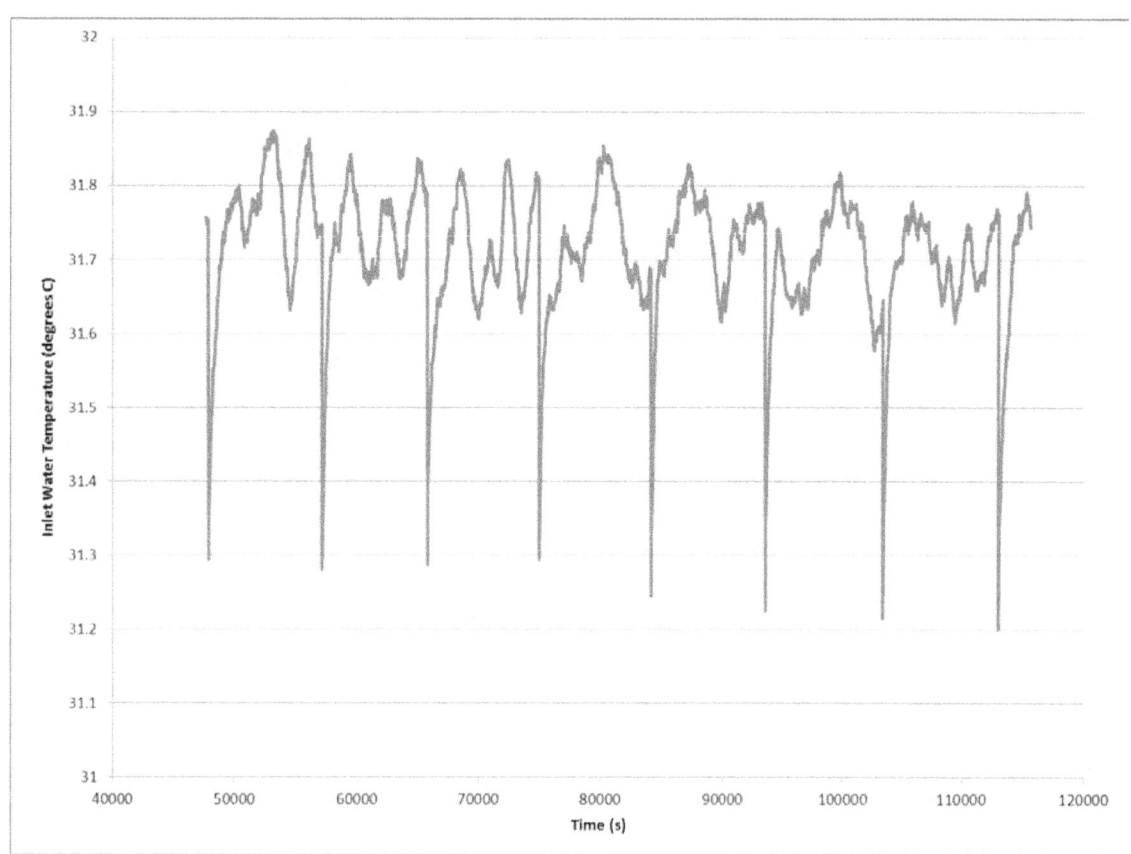

Figure 4.2 – Inlet Water Temperature

The water temperature was stable throughout the test period. However, there are 8 instances where the water temperature suddenly drops by nearly 0.5 °C. Note that the y-axis shows a small range in order to exaggerate the trend. This sudden change in measured temperature is caused by the water in the supply tube switching from stagnant to flowing for a short period of time. These points, therefore, indicate the time during which the refrigerator's solenoid valve was opened to fill the ice making molds. Figure 4.3 shows the water inlet temperature (red) on the same plot as the power draw (blue). Close examination shows a short spike in power draw aligned with each drop in water temperature. The increase in power draw is due to the operation of the solenoid valve, which consumes approximately 10 W for a period of less than 5 seconds. The power used by the auger is not obvious from the data; therefore our analysis of this unit's performance will center on the initiation of the solenoid as the marker for the beginning of each ice making cycle.

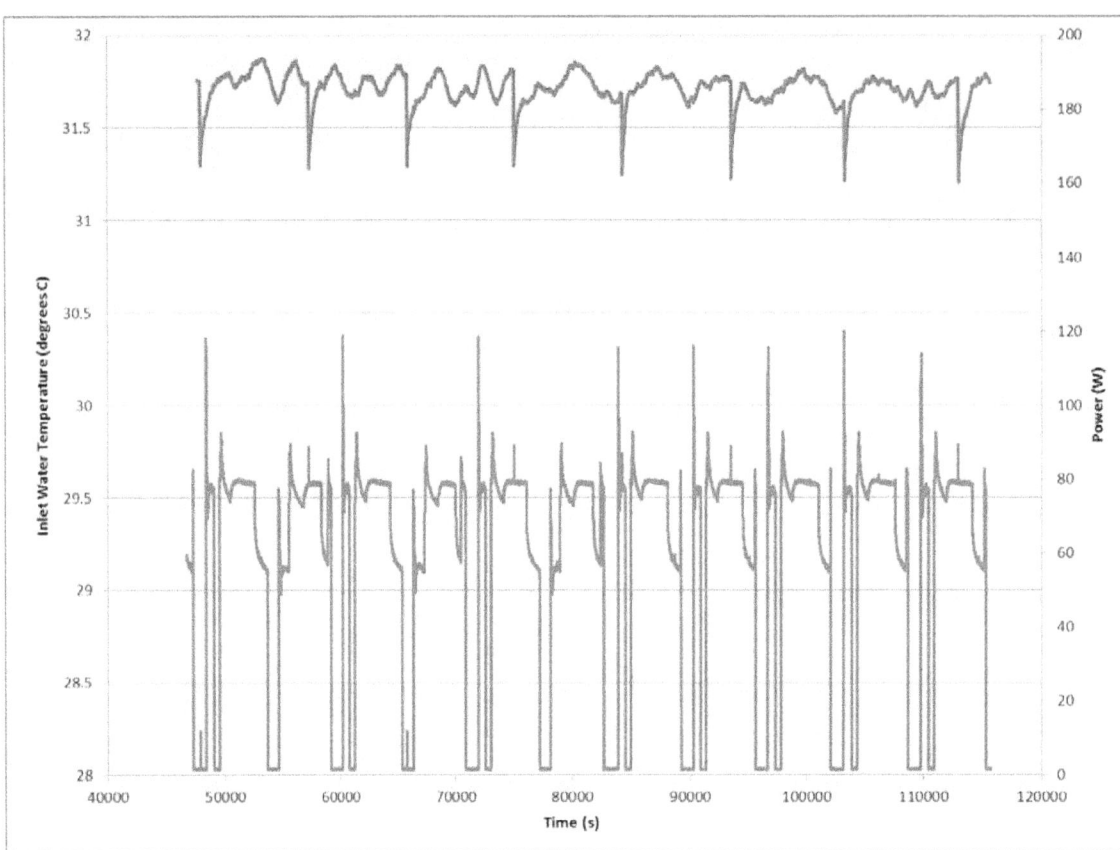

Figure 4.3 – Inlet Water Temperature and Power Draw at Mid Setting with Ice Production, Bottom Mount Unit

4.2.1 Mid Setting Results

For the first set of ice making tests we measured the energy consumption and compartment temperatures with the thermostats set to their median positions. The data set used from this experiment is shown graphically in Figure 4.4 and Figure 4.5 below. The figures below show the longest measured period during which the unit was operating under steady conditions and producing ice. There are two test periods marked on these figures; the test period outlined by the yellow rectangle consists of 7 whole ice making cycles while the test period marked by the purple rectangle consists of 9 whole compressor and temperature cycles. The test periods were marked this way based on the findings outlined in Chapter 3 of this report. The unit produced a total of 470 grams of ice during the test period marked by the orange rectangle; therefore our analysis assumes that each batch of ice harvested weighed 67.1 grams.

24

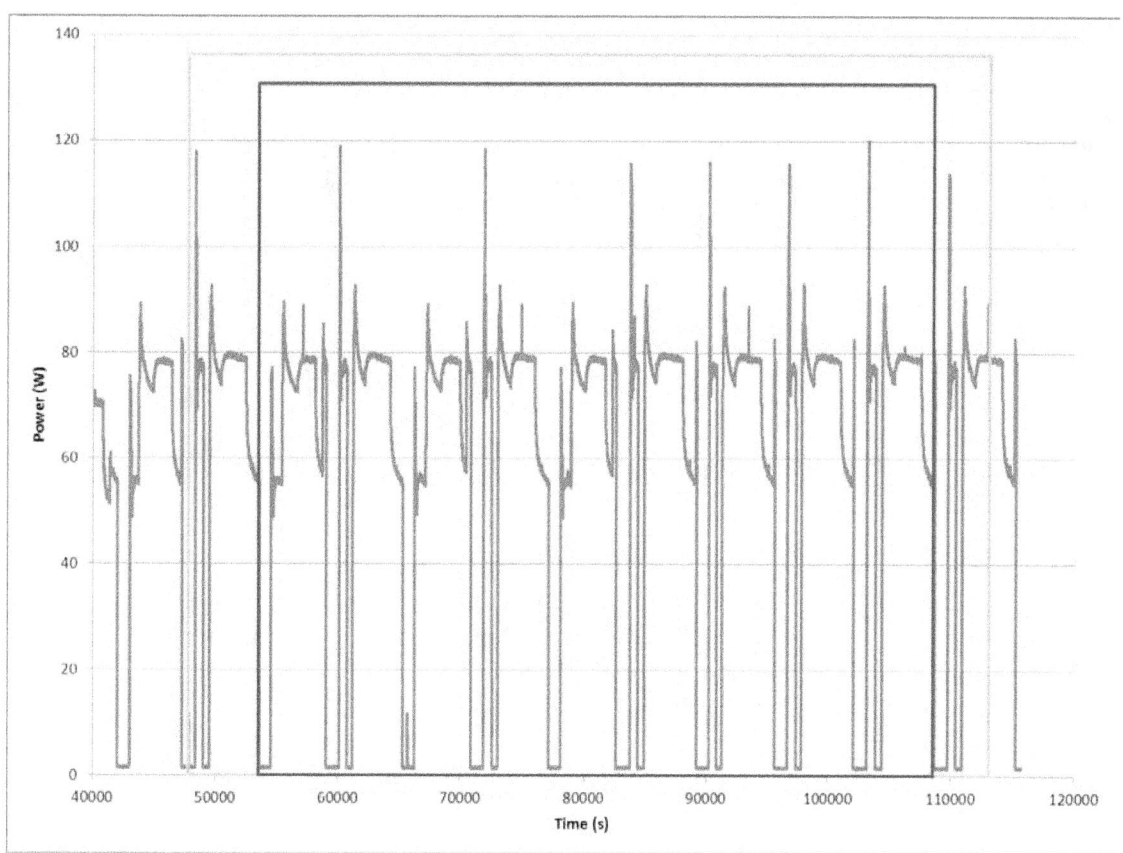

Figure 4.4 – Power at Mid Setting with Ice Production, Bottom Mount Unit

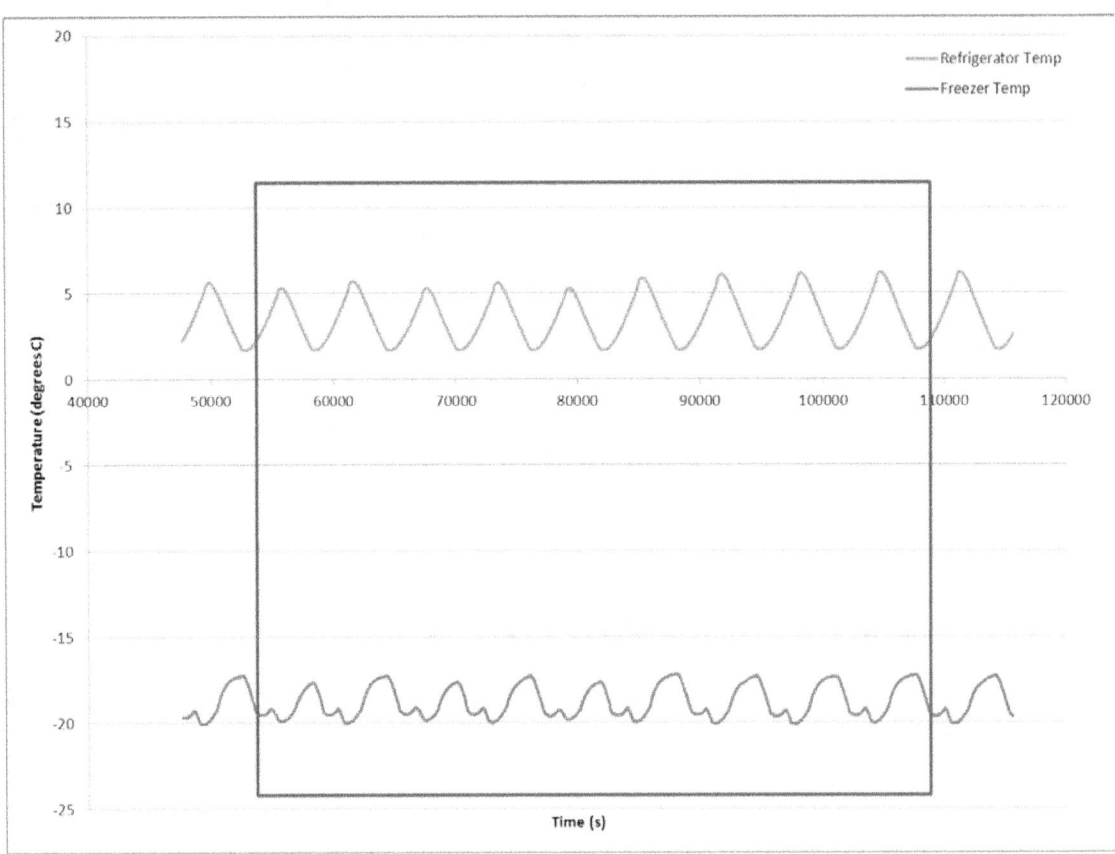

Figure 4.5 – Temperatures at Mid Setting with Ice Production, Bottom Mount Unit

Analysis of the data over the compressor cycle test period yielded the following results:

Cyclic operation time: 55121 seconds = 15:18:41
Measured refrigerator compartment temperature: (3.5 ± 0.1) ºC
Measured freezer compartment temperature: (-18.8 ± 0.1) ºC
Energy expended during the test period: (1045.8 ± 5.2) watt-hours

This yields an average power of (57.9 ± 0.3) watts over the duration of this test period. Figure 4.6 shows both the cumulative average power during the test period dictated by the compressor cycles (purple rectangle in the preceding figures), and the cumulative average energy per mass of ice produced during the test period dictated by the ice making cycles (yellow rectangle). The stability of the whole cycle values is apparent in this figure. The values of average power measured over whole compressor cycles stabilize to within 1 % of the final value of average power by the end of the 4th compressor cycle, total time of 23611 seconds (06:33:31).

The average energy per mass of ice produced is also very stable when examined over whole ice-making cycles. The end result over the whole data set is (2.225 ± 0.011) kWh of energy for each kilogram of ice produced. The values of average energy per mass of ice produced over whole ice-making cycles stabilize to within 2 % of the final value of average power by the end of the 4th ice making cycle, total time of 45612 seconds (12:40:12).

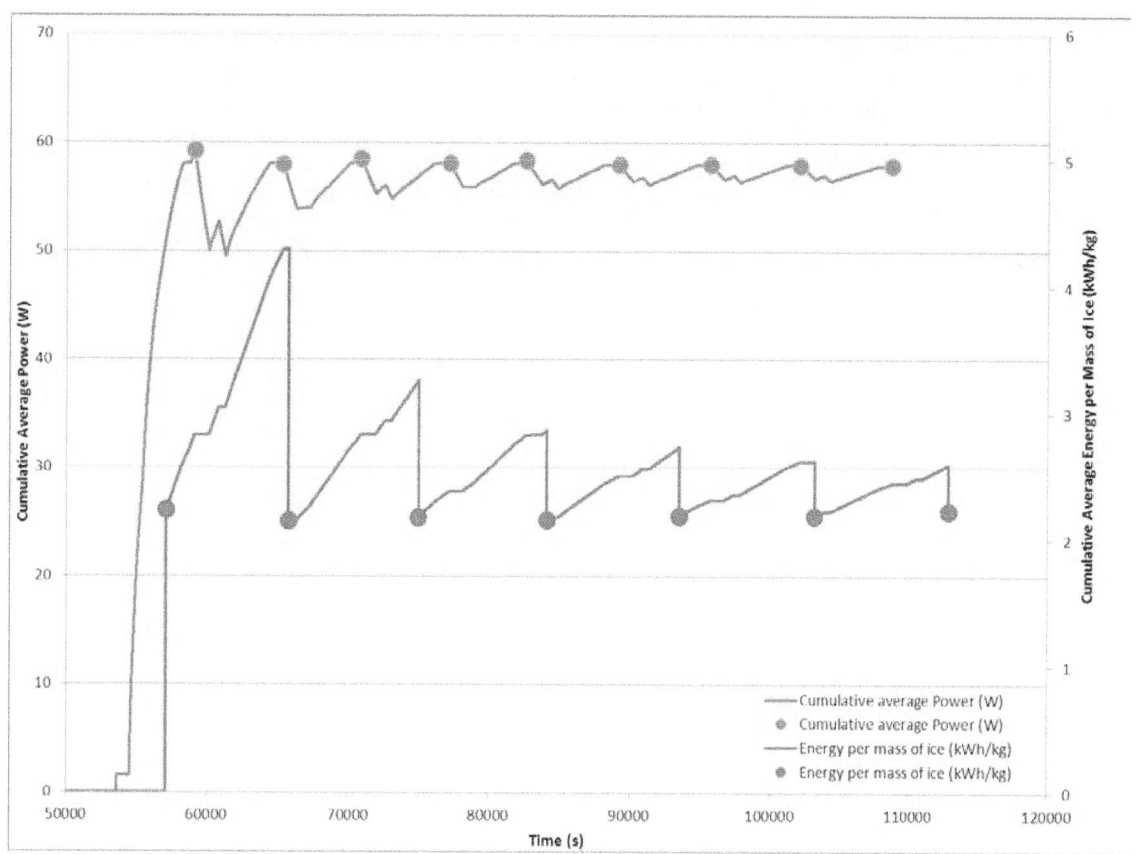

Figure 4.6 – Cumulative Average Power and Cumulative Average Energy per Mass of Ice at Mid Setting, Bottom Mount Unit

If we divide the average power of (57.9 ± 0.3) watts by this value, we obtain (0.0260 ± 0.0001) kg/h for the ice production rate.

Now that we have the ice production rate and the average power for this data set, we can determine the ice making energy using the simplest method; the method in which only data from the median thermostat settings is used. It is important to note that this refrigerator did not undergo any significant change in compartment temperature due to the operation of the icemaker; therefore these data sets should make a good basis for comparison.

We first need to determine the average power increase due to the operation of the icemaker by subtracting the average power obtained from the test at the same thermostat settings.

(57.9 ± 0.3) W - (53.3 ± 0.3) W = (4.6 ± 0.4) W

Then we can determine the icemaker energy consumption by dividing this value by the ice production rate.

27

(0.0046 ± 0.0004) kW $/ (0.0260 \pm 0.0001)$ kg/h $= \textbf{(0.177} \pm \textbf{0.015) kWh/kg}$

4.2.2 Warm Setting Results

The next set of measurements was used to examine the effects of interpolating the results from multiple ice making energy consumption tests. We set the thermostats for each compartment to their warmest setting and collected data. Figure 4.7 shows the power and temperatures measured during these tests. Again, the yellow rectangle marks the test period comprised of full ice making cycles, and the purple rectangle marks the test period comprised of full compressor cycles.

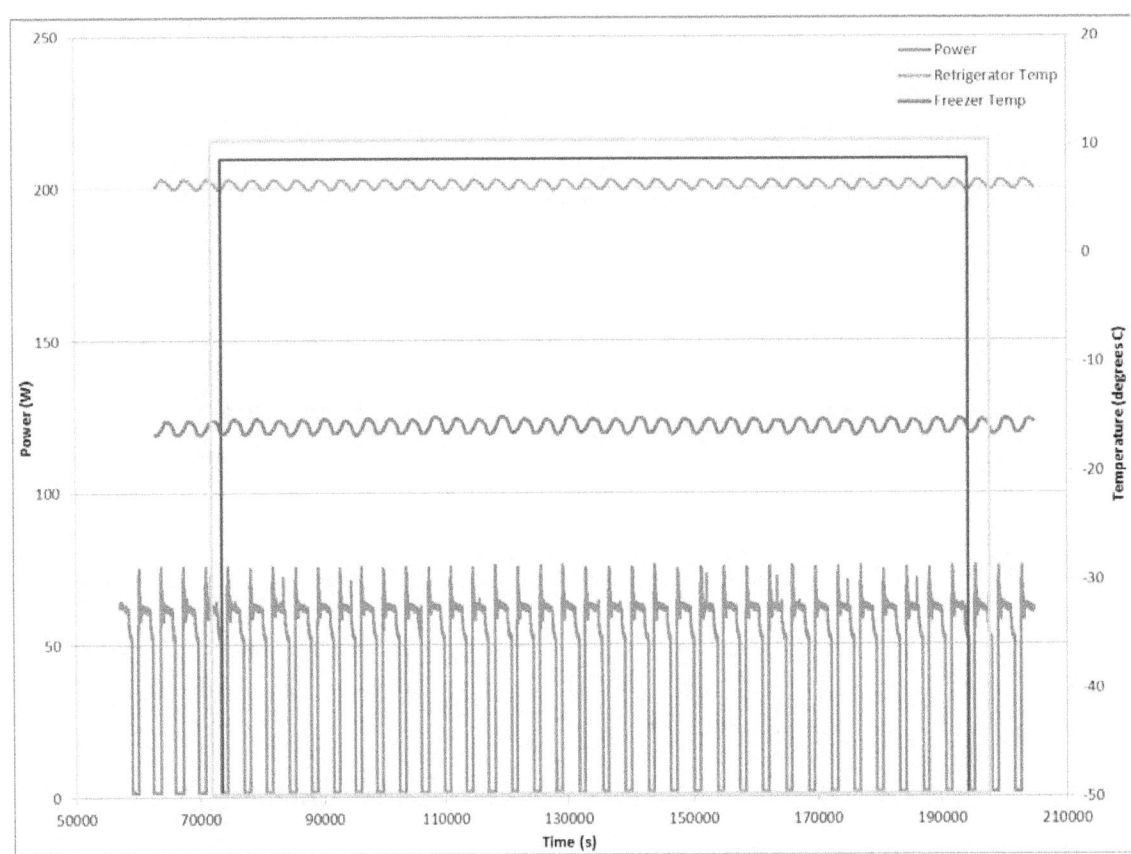

Figure 4.7 - Power and Temperatures at Warm Setting with Ice Production, Bottom Mount Unit

The yellow rectangle around the ice making test period includes15 ice harvest cycles and encompasses 125759 seconds (34:55:59) of clock time. The purple rectangle around the compressor based test period includes 33 compressor cycles and encompasses 120895 seconds (33:34:55) of clock time.

Figure 4.8 shows both the cumulative average power during the test period dictated by the compressor cycles and the cumulative average energy per mass of ice produced during the test period dictated by the ice making cycles. This figure illustrates the stability of these parameters when they are characterized over a whole number of cycles.

28

The end result of the cumulative average power over the whole data set is (44.6 ± 0.2) W. The values of average power measured over whole compressor cycles stabilize to within 1 % of the final value of average power by the end of the 10th compressor cycle, total time of 36359 seconds (10:05:59).

The average energy per mass of ice produced is also very stable when examined over whole ice making cycles. The end result over the whole data set is (2.050 ± 0.010) kWh of energy for each kilogram of ice produced. The values of average energy per mass of ice produced over whole ice making cycles stabilize to within 1 % of the final value of average power by the end of the 7th ice making cycle, total time of 80224 seconds (22:17:04); this parameter was always within 2 % of the end value throughout this data set.

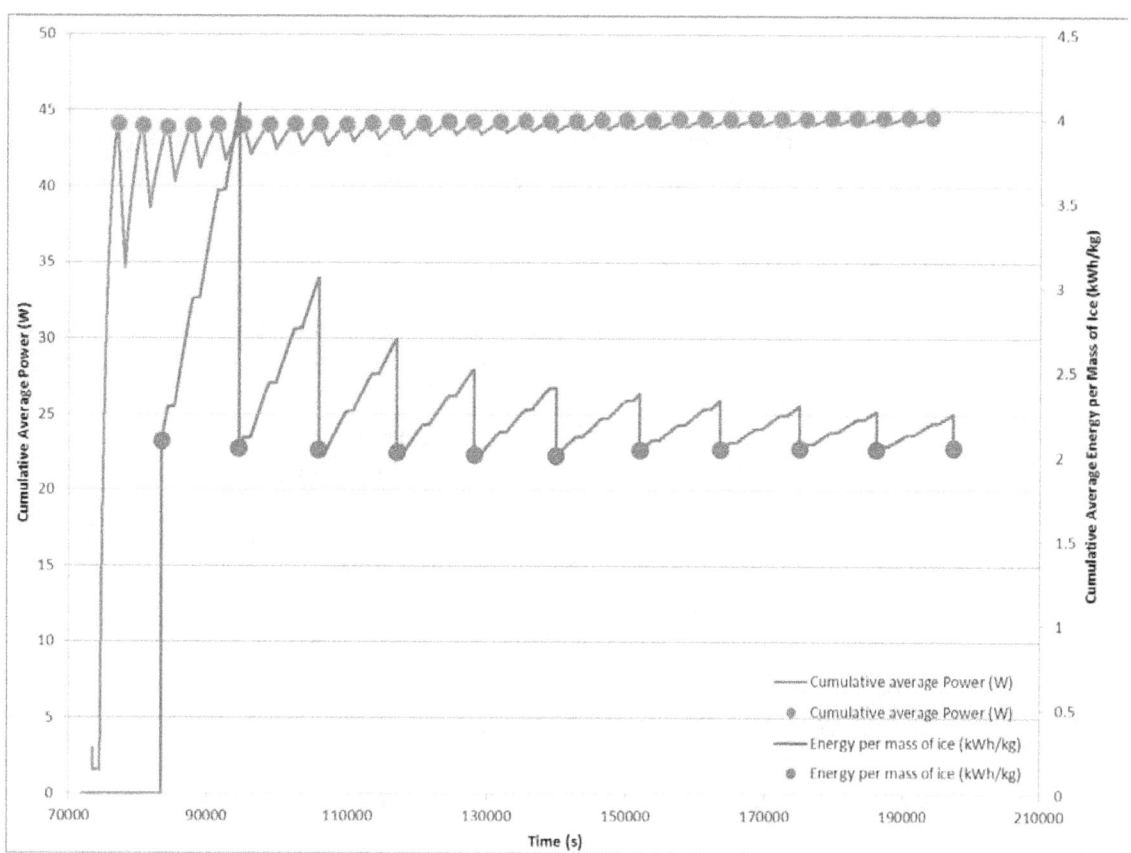

Figure 4.8 - Cumulative Average Power and Cumulative Average Energy per Mass of Ice at Warm Setting, Bottom Mount Unit

Measured refrigerator compartment temperature: (6.4 ± 0.1) °C
Measured freezer compartment temperature: (-15.9 ± 0.1) °C
Energy expended during the test period: (1497.8 ± 7.5) watt-hours
Average power: (44.6 ± 0.2) W

(0.0446 ± 0.0002) kW / (2.050 ± 0.010) kWh/kg = (0.0218 ± 0.0001) kg/h

Now that we have these parameters, it is again interesting (although not necessary) to calculate the increase in power consumption due to ice making energy at the warmest thermostat settings; again it is noted that this unit did not exhibit a significant change in cabinet temperature in response to the operation of the icemaker. In this case, we can examine the incremental power increase at these thermostat settings by subtracting the average power with and without the icemaker operational.

(44.6 ± 0.2) W - (41.6 ± 0.2) W = (3.4 ± 0.3) W

Then we can determine the icemaker energy consumption by dividing this value by the ice production rate.

(0.0034 ± 0.0003) kW / (0.0218 ± 0.0001) kg/h = **(0.156 ± 0.014) kWh/kg**

In this case, the difference is still small compared to the results of the median temperature setting test case when considering the measurement uncertainty.

4.2.3 Interpolating Ice-Making Results for Two Measurements
Now that we have the ice production rate, the average power, and the compartment temperatures for data sets acquired at the median and warmest settings, we can determine the ice making energy using a more complicated method.

This method is based on the values obtained via linear interpolation from the data sets to estimate the parameters at the target temperatures of 3.9 °C in the fresh food compartment and -17.8 °C in the frozen food compartment. The following results were obtained by interpolating the average power and ice production rate from the first two tests to the fresh food compartment temperatures while exceeding the frozen food compartment temperature; the temperature conditions were therefore 3.9 °C and -18.4 °C.

Average power = (56.1 ± 0.4) W
Ice Production Rate = (0.0254 ± 0.0001) kg/h

Using these values and those obtained earlier from the non-ice making tests, we can calculate the incremental power increase attributed to ice making.

(56.1 ± 0.4) W - (51.8 ± 0.3) W = (4.3 ± 0.5) W

The ice making energy consumption is therefore:

(0.0043 ± 0.0005) kW / (0.0254 ± 0.0001) kg/h = **(0.169 ± 0.020) kWh/kg**

4.2.4 Mixed Setting Results
For the sake of completeness, we will also examine a data set collected at a mixed thermostat setting. For this set, we set the freezer temperature to the coldest setting and

the refrigerator temperature to the median setting. Figure 4.9 shows the power and temperatures measured during these tests. Again, the yellow rectangle marks the test period comprised of full ice making cycles, and the purple rectangle marks the test period comprised of full compressor cycles.

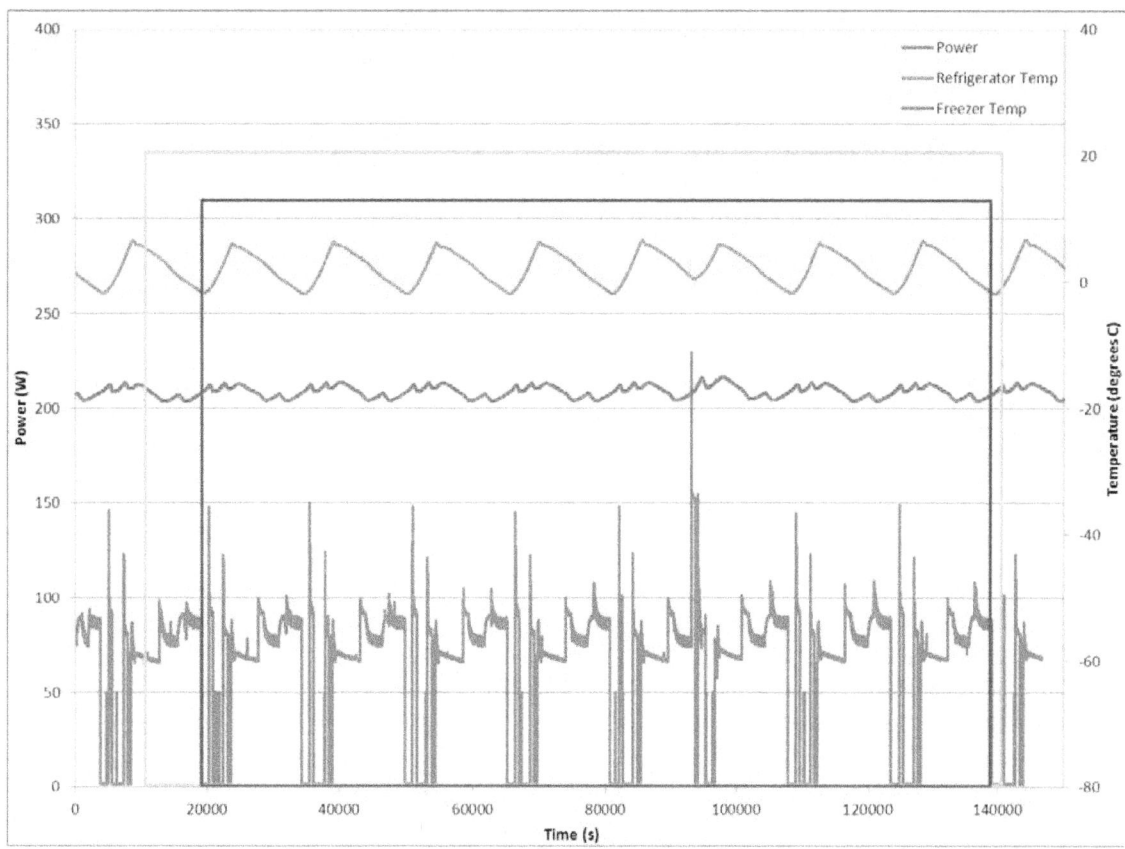

Figure 4.9 - Power and Temperatures at Mixed Setting with Ice Production, Bottom Mount Unit

Figure 4.10 shows both the cumulative average power during the test period dictated by the compressor cycles and the cumulative average energy per mass of ice produced during the test period dictated by the ice making cycles. The whole cycle value stability is also apparent in this figure, although the speed at which stability is attained is lower than seen in the other data sets due to the frequent shifting of the compressor. Using similar analyses, we can see that the cumulative average power is stable to with 1 % after the 5th compressor cycle or 76135 seconds (21:08:55). The end result of the cumulative average power over the whole data set is (64.2 ± 0.3) W.

The average energy per mass of ice produced is also very stable when examined over whole ice making cycles. The end result over the whole data set is (3.014 ± 0.015) kWh of energy for each kilogram of ice produced. The values of average energy per mass of ice produced over whole ice making cycles stabilize to within 2 % of the final value of average power by the end of the 10th ice making cycle, total time of 102085 seconds (28:21:25).

31

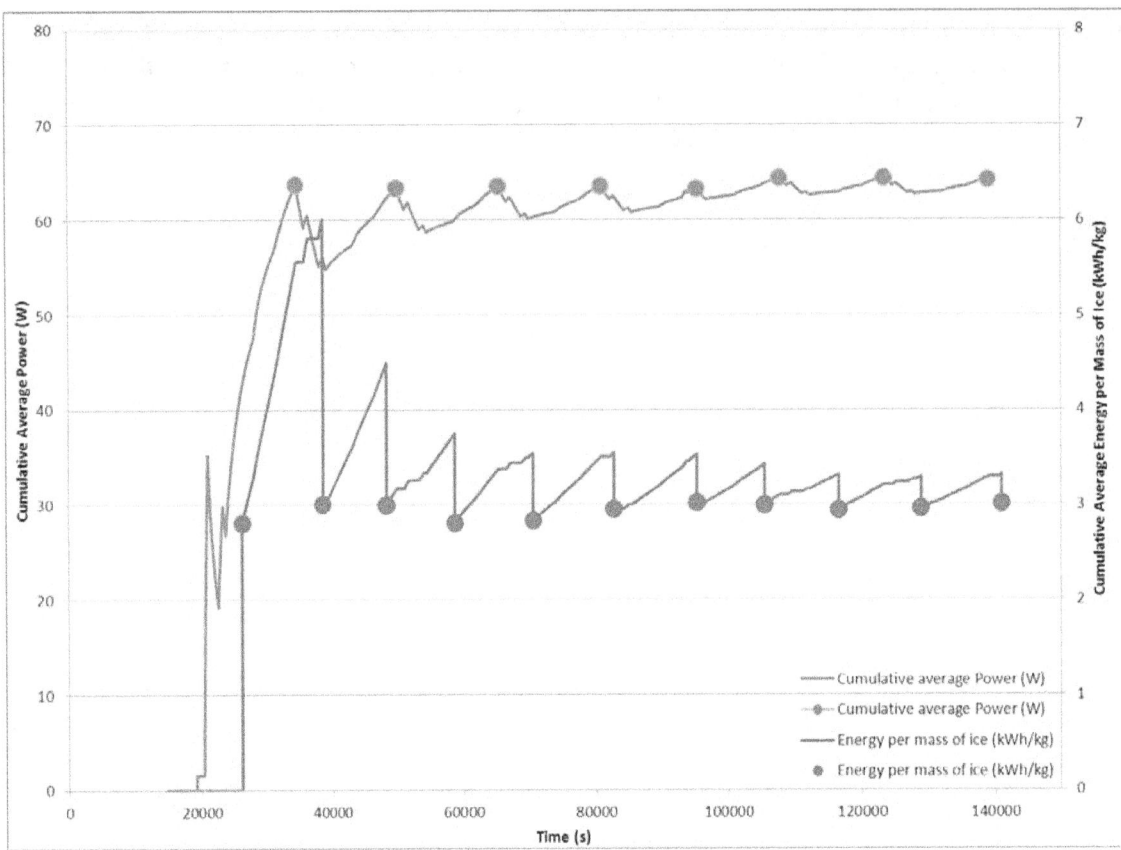

Figure 4.10 - Cumulative Average Power and Cumulative Average Energy per Mass of Ice at Mixed Setting, Bottom Mount Unit

Measured refrigerator compartment temperature: (2.2 ± 0.1) °C
Measured freezer compartment temperature: (-17.4 ± 0.1) °C
Energy expended during the test period: (2139.1 ± 10.7) watt-hours
Average power: (64.2 ± 0.3) W

Finally, the ice production rate can be calculated by dividing the average power by the energy to produce ice.

(0.0642 ± 0.0003) kW / (3.014 ± 0.015) kWh/kg = (0.0213 ± 0.0001) kg/h

We can again examine the differential energy consumption due to ice making at another single set of thermostat settings with this data set since the compartment temperatures did not change significantly in response to the operation of the icemaker. In this case, the incremental power increase is found by subtracting the average power with and without the icemaker operational.

(64.2 ± 0.3) W - (59.8 ± 0.3) W = (4.4 ± 0.4) W

Then we can determine the icemaker energy consumption by dividing this value by the ice production rate.

32

(0.0044 ± 0.0004) kW / (0.0213 ± 0.0001) kg/h = **(0.207 ± 0.019) kWh/kg**

Here, the difference is again larger than the results of the other single temperature setting test cases. However, it is important to note that the balance of cooling load between the refrigerator and freezer compartment during this test is vastly different from the other tests.

4.2.5 Interpolating Ice Making Results for Three Measurements

Lastly, we will assemble the three data sets using the triangular interpolation for both the power and ice production rates. First, we interpolate the three ice-making data sets to find the average power that would be realized if the exact target temperatures were realized. In this case, the calculated average power at a freezer temperature of -17.8 °C and a refrigerator temperature of 3.9 °C is (56.1 ± 0.5) W.

Next, we can apply the same calculation to the average power found during the ice making tests. If we apply the same calculation to the ice production rate, the calculated average ice production rate at a freezer temperature of -17.8 °C and a refrigerator temperature of 3.9 °C is (0.0240 ± 0.0003) kg/h.

Then, the differential power at the target temperatures is

(56.1 ± 0.5) W - (52.5 ± 0.5) W = (3.6 ± 0.7) W

And then the icemaker energy consumption is:

(0.0036 ± 0.0007) kW / (0.0240 ± 0.0003) kg/h = **(0.150 ± 0.02) kWh/kg**.

4.3 Summary

We analyzed six data sets from one refrigerator-freezer operating under various conditions. We used the same analysis techniques as in Chapter 3 of this report and tabulated the energy used to produce ice by 5 different methods of assembling this data. The table below summarizes the results.

Table 4.1: Summary of Test Results for Bottom Mount Unit with Twist Tray

Method	Compartment Temperatures Fridge and Freezer (°C)	Ice Making Energy (kWh/kg)
Single Point	$T_{FR} = 3.5$; $T_{FZ} = -18.8$	0.177 ± 0.015
	$T_{FR} = 6.4$; $T_{FZ} = -15.9$	0.156 ± 0.014
	$T_{FR} = 7.2$; $T_{FZ} = -17.4$	0.207 ± 0.019
2 point interpolation	$T_{FR} = 3.9$; $T_{FZ} < -17.8$	0.169 ± 0.020
3 point interpolation	$T_{FR} = 3.9$; $T_{FZ} = -17.8$	0.150 ± 0.020

Again we see that most of the values in the table are close to each other within the measurement uncertainty. It is, however, important to note that the measurement uncertainty is rather large for these experiments. This is due to the fact that the ice making energy is based on the difference between the power measurements taken while the unit is operating with and without ice production, and the difference in measured power is small.

There is one value that is a significant outlier, which was produced using the single data set method with mixed thermostatic setting. This method also produced an outlier result while analyzing the data from the previous unit because the unit's operation under these conditions was considerably different from the other data.

5: Review of Data from Prior Study

Through this study, we determined the steps that are necessary in order to measure the energy associated with the operation of an automatic icemaker. The nature of the operation dictates that it is necessary to examine different parameters using different test periods, which is unconventional for a test method of this type. Specifically, the average power drawn by the test unit should be measured over a whole number of compressor cycles because they are the overwhelming contributor to the measurement of average power. It is also necessary to determine the rate at which an icemaker produces ice, but this is not a parameter that can be measured during stable operation since it is highly variable. Instead, one must measure the average energy consumed during the production of a mass based quantity of ice and use it to calculate the production rate using the average power. The energy per mass of ice produced must be measured over a whole number of ice making cycles in order to provide a good measurement.

In consideration of these developments, we reexamined data published in a previous study (Yashar and Park, 2011), using the improved method. The first and second units examined in that study employed single speed compressors, and they are reexamined in this section. The third and fourth units employing variable speed compressors are not reexamined since their operation did not involve any compressor 'off' periods and the issue of nonsynchronous compressor and ice-making cycles is moot.

5.1 Top Mount Refrigerator Freezer Steady State Tests

The data in this section is taken from Yashar and Park (2011) and reiterated in this section for organizational purposes. It represents the steady state operation of this unit at various thermostat settings.

5.1.1 Top Mount Refrigerator Freezer Operating without Ice Production
5.1.1.1 Mid Setting Results

Refrigerator compartment temperature setting: Median
Freezer compartment temperature setting: Median

Steady state cyclic operation time: 10801 seconds = 03:00:01
Measured refrigerator compartment temperature: (4.0 ± 0.1) °C
Measured freezer compartment temperature: (-18.4 ± 0.1) °C
Energy expended during the test period: (164.2 ± 0.8) watt-hours

This yields an average steady state power of (54.7 ± 0.3) watts

5.1.1.2 Cold Setting Results

Refrigerator compartment temperature setting: Cold
Freezer compartment temperature setting: Cold

Steady state cyclic operation time: 12069 seconds = 03:21:09
Measured refrigerator compartment temperature: (3.0 ± 0.1) °C
Measured freezer compartment temperature: (-20.6 ± 0.1) °C
Energy expended during the test period: (199.2 ± 1.0) watt-hours

This yields an average steady state power of (59.4 ± 0.3) watts

5.1.1.3 Interpolating Steady State Results for Two Measurements
We interpolated the results of these measurements to determine the average power draw at the target temperatures. Using the refrigerator compartment temperature of 3.9 °C, the interpolation yields an average power of (55.2 ± 0.4) watts, while using the freezer compartment temperature of -17.8 °C yields an average power of (53.4 ± 0.3) watts. The higher of these two values is selected as the interpolation result.

5.1.2 Top Mount Refrigerator Freezer Operating with Ice Production
5.1.2.1 Mid Setting Results
This section examines the ice making energy using the outlined methods. The data taken from the first measurement, at the median temperature setting, is discussed first.
Figure 5.1 shows the cumulative average power draw (blue) over whole compressor cycles and the cumulative average energy per mass of ice produced (red) over whole ice making cycles. Each of these parameters is calculated relative to a different start point and the circles mark the end of cycle values. It is clear from this figure that each of these parameters is fairly stable when only examining the end of cycle values. The average power parameter stays within 1 % of its final value after the 3rd compressor cycle, a total of 12266 seconds (3:24:26) of clock time; while the average energy parameter stays within 1 % of its final value after the 5th batch of ice is produced, a total of 27934 seconds (7:45:34) of clock time.

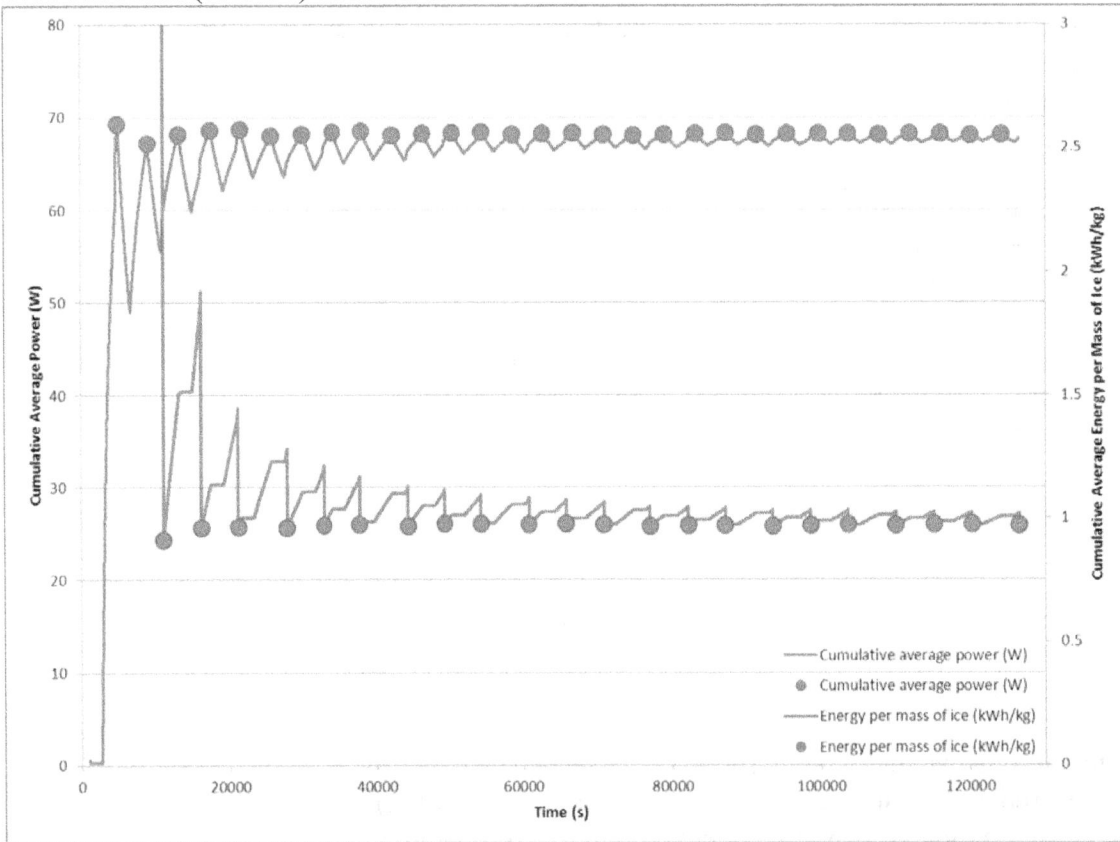

Figure 5.1 – Cumulative Average Power and Cumulative Average Energy per Mass of Ice at Mid Setting, Top Mount Unit

The measured values are listed below.
Measured refrigerator compartment temperature: (3.7 ± 0.1) °C
Measured freezer compartment temperature: (-16.2 ± 0.1) °C
Average steady state power: (68.2 ± 0.3) watts
Average energy per mass of ice produced: (0.972 ± 0.005) kWh/kg

The production rate is calculated by dividing the average power by the average energy per mass of ice produced.

(68.2 ± 0.3) W / (972 ± 5) Wh/kg = (0.0702 ± 0.0005) kg/hr

Next, we will compare the results of this test to the results of the energy consumption test at the same thermostat setting without ice production. The difference in steady state power is calculated below.

(68.2 ± 0.3) W - (54.7 ± 0.3) W = (13.5 ± 0.4) W

Then we can determine the icemaker energy consumption by dividing this value by the ice production rate.

(0.0135 ± 0.0004) kW / (0.0702 ± 0.0005) kg/h = **(0.192 ± 0.005) kWh/kg**

It is important to note that this unit exhibited a large increase in freezer temperature in response to the initiation of the ice making process, therefore the icemaker energy consumption is much lower than what would be obtained if the freezer temperature did not shift when the icemaker began producing new ice.

5.1.2.2 Cold Setting Results
Next, we will examine the data taken from the second measurement, at the coldest temperature setting. Figure 5.2 shows the cumulative average power draw (blue) over whole compressor cycles and the cumulative average energy per mass of ice produced (red) over whole ice making cycles. The figure shows that the average power parameter stays within 1 % of its final value after the 3rd compressor cycle, a total of 11366 seconds (3:09:26) of clock time; while the average energy parameter stays within 1 % of its final value after the 7th batch of ice is produced, a total of 30934seconds (8:35:34) of clock time.

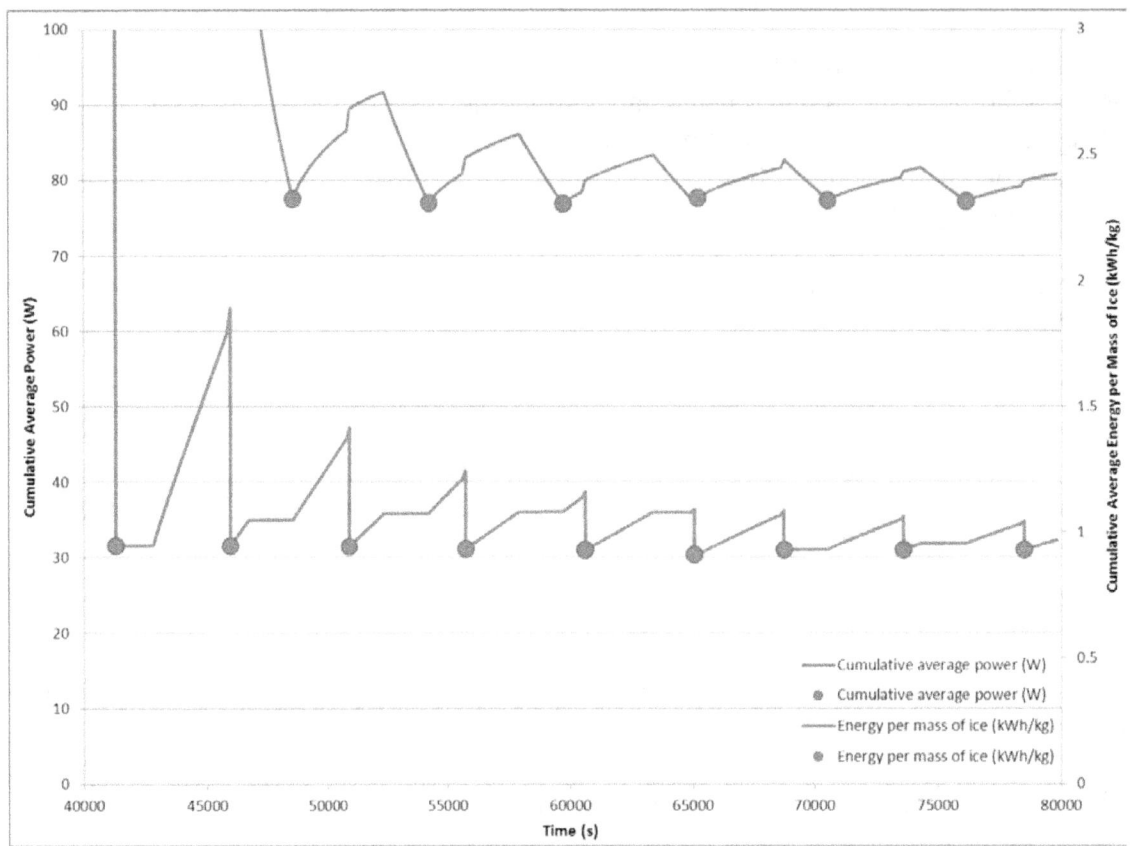

Figure 5.2 – Cumulative Average Power and Cumulative Average Energy per Mass of Ice at Cold Setting, Top Mount Unit

The measured values are listed below.
Measured refrigerator compartment temperature: (-2.4 ± 0.1) °C
Measured freezer compartment temperature: (-18.1 ± 0.1) °C
Average steady state power: (77.2 ± 0.4) watts
Average energy per mass of ice produced: (0.930 ± 0.005) kWh/kg

The production rate is calculated by dividing the average power by the average energy per mass of ice produced.

(77.2 ± 0.4) W / (930 ± 5) Wh/kg = (0.0830 ± 0.0006) kg/hr

Next, we will compare the results of this test to the results of the energy consumption test at the same thermostat setting without ice production. The difference in steady state power is calculated below.

(77.2 ± 0.4) W - (59.4 ± 0.3) W = (17.8 ± 0.5) W

Then we can determine the icemaker energy consumption by dividing this value by the ice production rate.

(0.0178 ± 0.0005) kW / (0.0830 ± 0.0006) kg/h = **(0.214 ± 0.006) kWh/kg**

38

Similar to the previous data set, the freezer temperature significantly increased in response to the initiation of the ice-making process; therefore this value of ice-making energy consumption is also smaller than that which would have been obtained if the temperature in the unit did not change.

5.1.2.3 Interpolating Ice Making Results for Two Measurements
Now that we have the ice production rate, the average power, and the compartment temperatures for data sets acquired at the median and coldest settings, we can determine the ice making energy using two point interpolation. The following results were obtained for the average power and ice production rate at the more stringent of the two target temperatures of 3.9 °C and -17.8 °C:

Average power = (75.8 ± 0.6) W
Ice production rate = (0.0810 ± 0.0009) kg/h

Using these values and those obtained earlier from the non-ice making tests, we can calculate the incremental power increase attributed to ice making.

(75.8 ± 0.6) W - (53.4 ± 0.3) W = (22.4 ± 0.7) W

The ice making energy consumption calculated by linearly interpolating the results of two measurements is therefore:

(0.0224 ± 0.0005) kW / (0.0810 ± 0.0009) kg/h = **(0.277 ± 0.007) kWh/kg**

This value is considerably different from the single point ice making energy values. This is because the ice-making temperatures used for each data point do not match up well with the corresponding non ice-making temperature data. Even the interpolated values calculated from the interpolation between the coldest and median settings occur at a point far from the target due to the large temperature shift caused by the icemaker operation. The interpolation results met the frozen food compartment temperature of -17.8 °C while at a fresh food compartment temperature of -1.4 °C; for this reason a third data point at a mixed setting was taken during the original study.

5.1.2.4 Mixed Setting Results
The results of this third data point are examined next. Figure 5.3 shows the cumulative average power draw (blue) over whole compressor cycles and the cumulative average energy per mass of ice produced (red) over whole ice making cycles. The figure shows that the average power parameter stays within 1 % of its final value after the 3rd compressor cycle, a total of 16907 seconds (4:41:47) of clock time; while the average energy parameter stays within 1 % of its final value after the 6th batch of ice is produced, a total of 28708 seconds (7:58:28) of clock time.

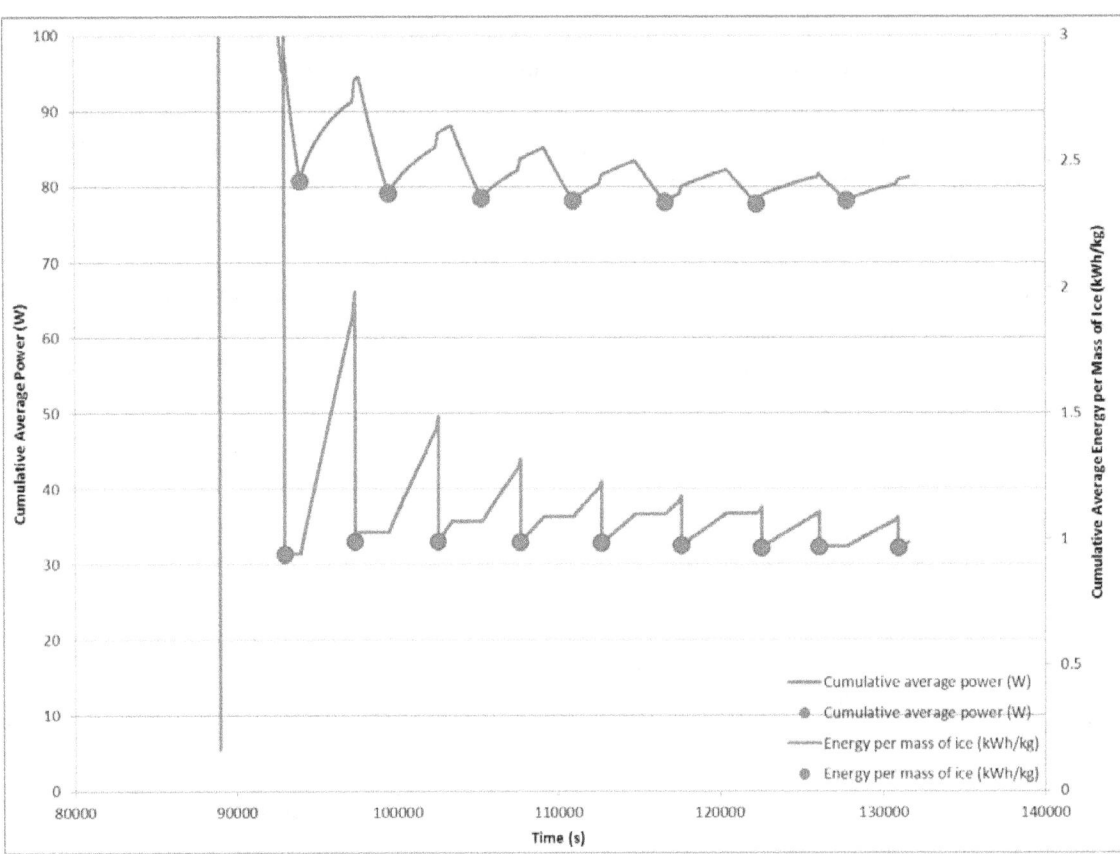

Figure 5.3 – Cumulative Average Power and Cumulative Average Energy per Mass of Ice at Mixed Setting, Top Mount Unit

The measured values are listed below.
Measured refrigerator compartment temperature: (4.7 ± 0.1) °C
Measured freezer compartment temperature: (-17.2 ± 0.1) °C
Average steady state power: (78.1 ± 0.4) watts
Average energy per mass of ice produced: (0.967 ± 0.005) kWh/kg

The production rate is calculated by dividing the average power by the average energy per mass of ice produced.

(78.1 ± 0.4) W / (967 ± 5) Wh/kg = (0.0808 ± 0.0004) kg/hr

5.1.2.5 Interpolating Ice Making Results for Three Measurements
Now we can determine the ice making energy using three point interpolation. The following results were obtained for the average power and ice production rate at the target temperatures of 3.9 °C and -17.8 °C:

Average power = (82.3 ± 2.3) W
Ice production rate = (0.0859 ± 0.002) kg/h

Using these values and those obtained earlier from the non-ice making tests, we can calculate the incremental power increase attributed to ice making.

40

(82.3 ± 2.3) W - (55.2 ± 0.4) W = (27.1 ± 2.3) W

The ice making energy consumption is therefore:

(0.0271 ± 0.0023) kW / (0.0859 ± 0.002) kg/h = **(0.315 ± 0.028) kWh/kg**

5.1.3 Top Mount Refrigerator Freezer Summary
The table below summarizes the results of all of the calculations.

Table 5.1: Summary of Test Results for Top Mount Unit

Method	Compartment Temperatures Fridge and Freezer (°C)	Ice Making Energy (kWh/kg)
Single Point	T_{FR} = 3.7; T_{FZ} = -16.2	0.192 ± 0.005
	T_{FR} = -2.4; T_{FZ} = -18.1	0.214 ± 0.006
2 point interpolation	T_{FR} << 3.9; T_{FZ} = -17.8	0.277 ± 0.007
3 point interpolation	T_{FR} = 3.9; T_{FZ} = -17.8	0.315 ± 0.028

The two most interesting values in this table are the values listed in the top and bottom rows. Both values have considerable significance. The single point value at the median setting is that which would be expected if the thermostat position were held constant. The single point value is most likely to occur during normal operation; however, there is no standardized condition or even limit to the extent that the compartment temperatures may deviate from the norm during the measurement. The interpolation values are those which would be expected if the compartment temperatures were fixed to a common target. However, the 2 point interpolation value is based on a refrigerator temperature that is much colder than the target therefore it is not a very good representation. While the 3 point interpolation is properly fixed to the target temperatures, it is important to note that the measurement uncertainty for the 3 point interpolation result is large. The uncertainty of the 3 point method is influenced by the proximity of the measured temperature data to the target, which cannot be easily controlled; therefore large uncertainty may be common for such measurements. It is also important to note that the 3 point interpolation method requires a significantly greater amount of test time due to the number of data points.

5.2 Side-by-Side Refrigerator Freezer Steady State Tests
The data in this section is taken from Chapter 4 of NIST Technical Note 1697 and reiterated in this section for organizational purposes. It represents the steady state operation of this unit at various thermostat settings.

5.2.1 Side-by-Side Refrigerator Freezer Operating without Ice Production
5.2.1.1 Mid Setting Results
Refrigerator compartment temperature setting: Median
Freezer compartment temperature setting: Median

Steady state cyclic operation time: 14290 seconds = 03:58:10

Measured refrigerator compartment temperature: (4.2 ± 0.1) °C
Measured freezer compartment temperature: (-17.7 ± 0.1) °C
Energy expended during the test period: (401.1 ± 2.0) watt-hours

This yields an average steady state power of (101.1 ± 0.5) watts.

5.2.1.2 Cold Setting Results
Refrigerator compartment temperature setting: Cold
Freezer compartment temperature setting: Cold

Steady state cyclic operation time: 20715 seconds = 05:45:15
Measured refrigerator compartment temperature: (-2.0 ± 0.1) °C
Measured freezer compartment temperature: (-21.5 ± 0.1) °C
Energy expended during the test period: (729.9 ± 3.6) watt-hours

This yields an average steady state power of (126.8 ± 0.6) watts.

5.2.1.3 Interpolating Steady State Results for Two Measurements
We interpolated the results of these measurements to determine the average power draw
at the target temperatures. Using the refrigerator compartment temperature of 3.9 °C, the
interpolation yields an average power of (102.3 ± 0.4) watts, while using the freezer
compartment temperature of -17.8 °C yields an average power of (101.8 ± 0.7) watts.
The higher of these two values is selected as the interpolation result.

5.2.2 Side-by-Side Refrigerator Freezer Operating with Ice Production
5.2.2.1 Mid Setting Results
This section examines the ice making energy using the outlined methods. Figure 5.4
shows the cumulative average power draw (blue) over whole compressor cycles and the
cumulative average energy per mass of ice produced (red) over whole ice making cycles
for the data taken at the median temperature settings. The average power parameter stays
within 1 % of its final value after the 1st compressor cycle, a total of 6467 seconds
(1:47:47) of clock time; while the average energy parameter stays within 1 % of its final
value after the 5th batch of ice is produced, a total of 20328 seconds (5:38:48) of clock
time.

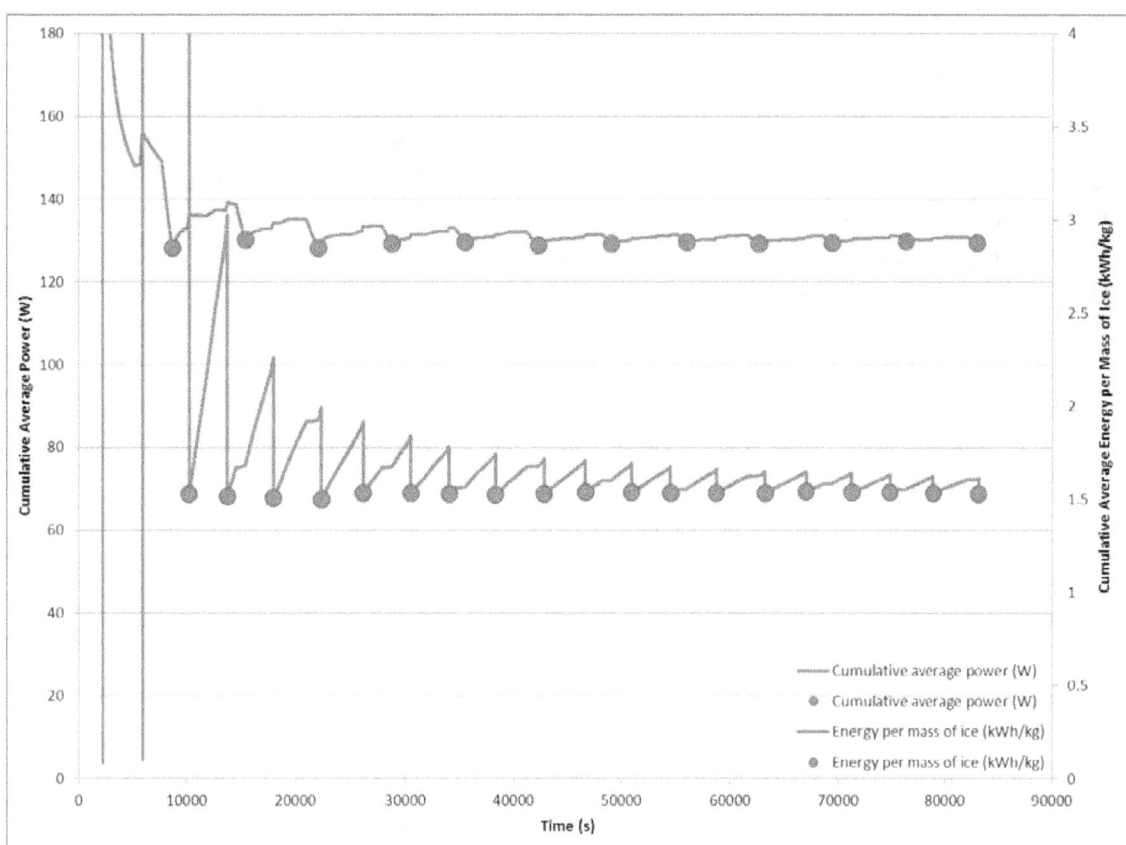

Figure 5.4 - Cumulative Average Power and Cumulative Average Energy per Mass of Ice at Mid Setting, Side-by-Side Unit

The measured values are listed below.
Measured refrigerator compartment temperature: (3.1 ± 0.1) °C
Measured freezer compartment temperature: (-17.7 ± 0.1) °C
Average steady state power: (129.5 ± 0.6) watts
Average energy per mass of ice produced: (1.526 ± 0.008) kWh/kg

The production rate is calculated by dividing the average power by the average energy per mass of ice produced.

(129.5 ± 0.6) W / (1526 ± 8) Wh/kg = (0.0849 ± 0.0006) kg/hr

Next, we will compare the results of this test to the results of the energy consumption test at the same thermostat setting without ice production. The difference in steady state power is calculated below.

(129.5 ± 0.6) W - (101.1 ± 0.5) W = (28.4 ± 0.8) W

Then we can determine the icemaker energy consumption by dividing this value by the ice production rate.

(0.0284 ± 0.0008) kW / (0.0849 ± 0.0006) kg/h = **(0.335 ± 0.010) kWh/kg**

43

5.2.2.2 Cold Setting Results

In the previous study, we also performed measurements with the thermostats set to their coldest settings. This caused the compressor to operate 100 % of the time, therefore there was no compressor cycling. The unit did, however, exhibit temperature cycles because the cooling load was periodically shifted between the refrigerator and freezer compartments; in light of this we used the temperature cycles to define the cyclic periods. Figure 5.5 shows the operation of this unit under these conditions.

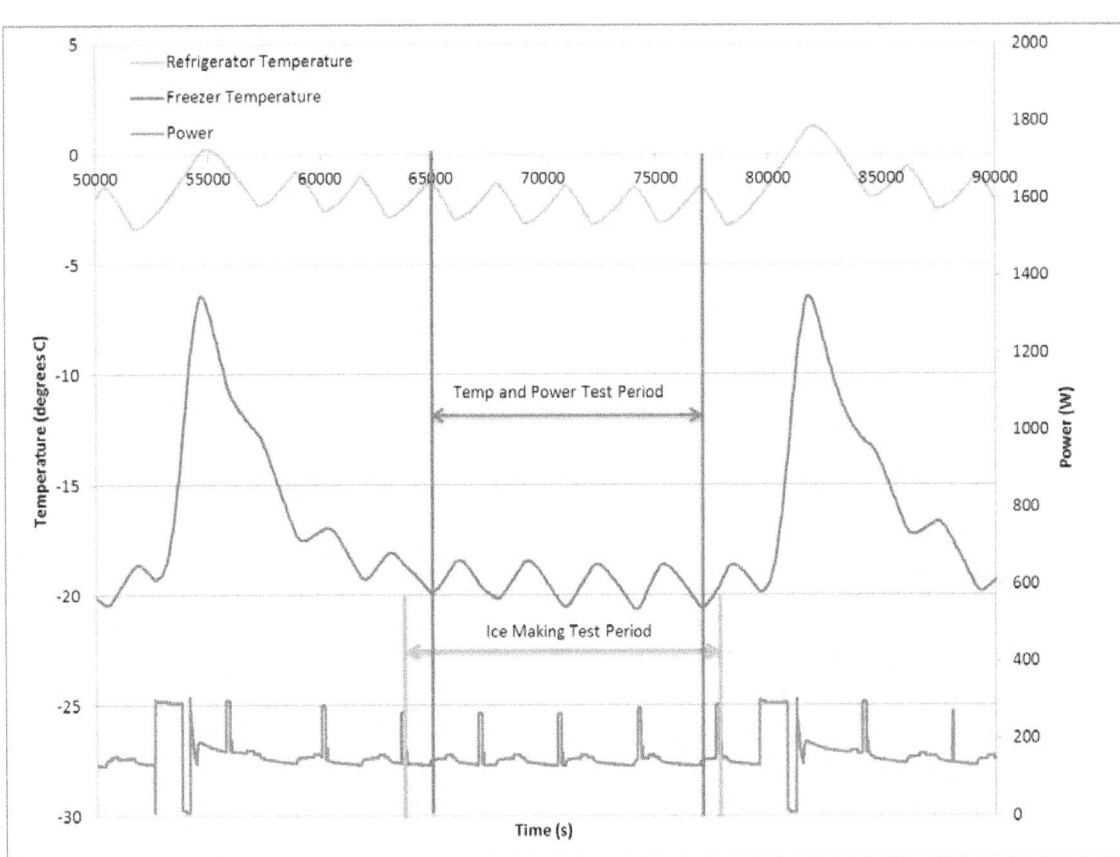

Figure 5.5 – Power and Temperatures at Mixed Setting with Ice Production, Side-by-Side Unit

Figure 5.6 shows the data organized to show the cumulative average power draw and the cumulative average energy per mass of ice produced. Overall, the cumulative average power does not vary much with time since the compressor does not turn off. The ice harvest heaters have a small influence on the cumulative average power draw, however the significance is small. When examining this parameter over whole temperature cycles, the influence has less than a 1 % impact after the 3rd temperature cycle or 9143 seconds (2:32:23) of clock time. The cumulative average energy per mass of ice produced is also very stable and does not deviate from the end value by more than 1 % after the 2nd ice making cycle, or 7007 seconds (1:56:47) of clock time.

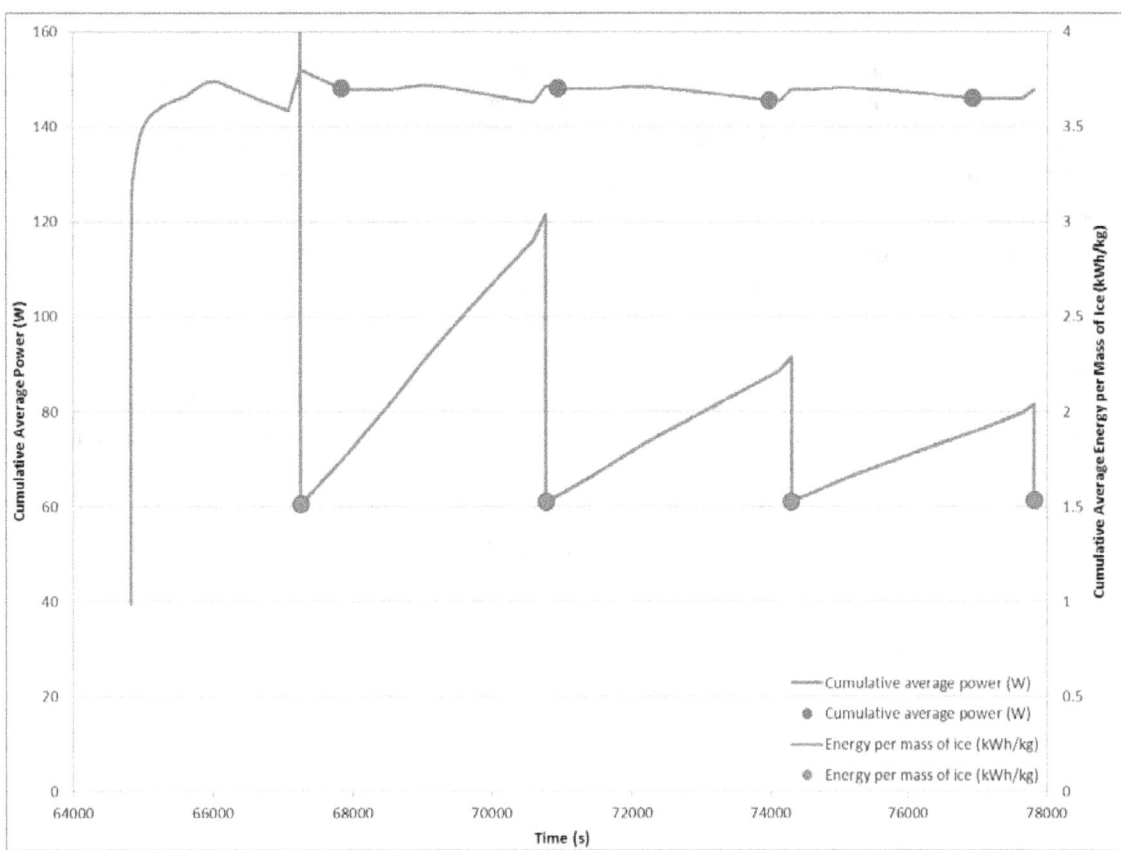

Figure 5.6 – Cumulative Average Power and Cumulative Average Energy per Mass of Ice at Cold Setting, Side-by-Side Unit

The measured values are listed below.
Measured refrigerator compartment temperature: (-2.3 ± 0.1) °C
Measured freezer compartment temperature: (-19.4 ± 0.1) °C
Average steady state power: (146.1 ± 0.7) watts
Average energy per mass of ice produced: (1.532 ± 0.008) kWh/kg

The production rate is calculated by dividing the average power by the average energy per mass of ice produced.

(146.1 ± 0.7) W / (1532 ± 5) Wh/kg = (0.0954 ± 0.0006) kg/hr

Next, we compared the results of this test to the results of the energy consumption test at the same thermostat setting without ice production. The difference in steady state power is calculated below.

(146.1 ± 0.7) W - (126.8 ± 0.6) W = (19.3 ± 0.9) W

Then we determined the icemaker energy consumption at the coldest thermostat setting by dividing this value by the ice production rate.

(0.0193 ± 0.0009) kW / (0.0954 ± 0.0006) kg/h = **(0.202 ± 0.010) kWh/kg**

45

It is important to note that the results acquired from this data set showed a significant change in freezer temperature in response to ice production, which is the reason that the icemaker energy consumption is significantly lower than the previous calculation at the median setting. The data from the median setting did not exhibit the same temperature change, therefore it is likely that the unit was working at its capacity limit to produce ice and maintain the cabinet temperatures at the coldest settings. This also makes sense since the compressor did not switch off during this data set.

5.2.2.3 Interpolating Steady State Results for Two Measurements
Now that we have the ice production rate, the average power, and the compartment temperatures for data sets acquired at the median and coldest settings, we can determine the ice making energy using two point interpolation. The following results were obtained by interpolating the average power and ice production rate from the first two tests to the frozen food compartment temperatures while exceeding the fresh food compartment temperature; the temperature conditions were therefore 2.8 °C and -17.8 °C.

Average power = (130.5 ± 0.9) W
Ice production rate = (0.0855 ± 0.0006) kg/h

Using these values and those obtained earlier from the non-ice making tests, we can calculate the incremental power increase attributed to ice making.

(130.5 ± 0.9) W - (102.3 ± 0.4) W = (28.2 ± 1.0) W

The ice making energy consumption is therefore:

(0.0282 ± 0.0010) kW / (0.0855 ± 0.0006) kg/h = **(0.330 ± 0.012) kWh/kg**

5.2.3 Side-by-Side Refrigerator Freezer Summary
The table below summarizes the results of all of the calculations.

Table 5.2: Summary of Test Results for Side-by-Side Unit

Method	Compartment Temperatures Refrigerator and Freezer (°C)	Ice Making Energy (kWh/kg)
Single Point	$T_{FR} = 3.1$; $T_{FZ} = -17.7$	0.335 ± 0.010
	$T_{FR} = -2.3$; $T_{FZ} = -19.4$	0.202 ± 0.010
2 point interpolation	$T_{FR} < 3.9$; $T_{FZ} = -17.8$	0.330 ± 0.012

When operating at the median thermostat settings, this unit did not realize any significant change in cabinet temperature when the icemaker was rendered operative. When operated at the coldest setting, however, the freezer temperature increased by nearly 4 K. Therefore, the first single point value of ice making energy was calculated comparing operation at (more or less) the same temperatures while the second value was calculated by comparing operation at different temperatures. The second value is significantly lower because the unit was maintaining a warmer freezer temperature while making ice than it was without ice production. The interpolated value accounts for temperature variation and is therefore similar to the first single point value.

6: Summary

Two refrigerator-freezers with automatic icemakers were tested in this study. In addition, data from two units studied in a previous work was also re-examined and included in this study. Data from these four units was used to evaluate procedure options for determining energy consumption due to ice production. Characterizing the energy consumption due to ice production requires three important parameters which include the average power drawn by the unit with and without ice production and the rate at which the icemaker produces ice.

Icemakers typically operate on a cycle that is independent of the unit's compressor cycle; therefore, there is a lot of difficulty separating out the icemaker's contribution to overall energy use without substantial truncation error. Through analysis of several sets of test data, however, we were able to determine a repeatable and reproducible method of characterizing the ice making energy consumption. The important findings of this study are:

- The power consumed by the refrigerator during stable cyclic operation must be determined from data that includes a whole number of compressor cycles, whether it is or isn't operating the icemaker.
- The rate at which an automatic icemaker produces ice is highly variable due to the mechanisms that drive ice production; therefore it is not possible to directly measure this quantity with good confidence.
- When the icemaker is actively producing batches of ice, the energy consumed during the production of each batch is measurable quantity that is stable.
- The rate at which an automatic icemaker produces ice may be calculated by dividing the unit's average power by the average energy per batch of ice produced.

Therefore, the method developed in this study is somewhat unconventional in that it requires parameters to be extracted from two different data subsets within a test period; specifically, the power must be measured over a whole number of compressor cycles within the data set and the energy per batch of ice must be measured over a whole number of ice making cycles within the data set.

In total, data from four units consisting of different icemaker technologies were examined, and this study showed that a robust test method can be successfully applied to this range of designs. Interestingly, the results of all of the measurements showed that the most efficient icemaker tested consumed less than half of the energy that was consumed by the least efficient unit, indicating that there is considerable performance diversity in the current market.

6.1 Method to Measure the Energy Consumption of Automatic Icemakers

The following steps are recommended to determine the amount of energy use attributed to the operation of an automatic icemaker.

1) Determine the average power [W] drawn by the unit during steady operation without ice production. For a unit with a cycling compressor, this parameter must be characterized over a whole number of complete compressor cycles.

2) Determine the average power [W] drawn by the unit during steady operation with ice production. For a unit with a cycling compressor, this parameter must also be characterized over a whole number of complete compressor cycles. This is because the average power is a much stronger function of the compressor operation than the icemaker operation. If, and only if, the compressor does not cycle during the test period, then this parameter should be characterized over a whole number of ice making cycles.

3) Determine the amount of energy consumed per batch [Wh/batch] (or mass based quantity [Wh/kg or Wh/lb$_m$]) of ice produced while maintaining cabinet temperatures. This parameter must be characterized over a whole number of ice making cycles and can be found by dividing the accumulated energy over a whole number of ice making cycles by the number of ice making cycles (or equivalently the mass of ice produced during that test period).

4) Calculate the average production rate of ice [batch/time or mass/time] by dividing the energy per batch (or per mass) found in step (3) by the average power found in step (2).

5) Calculate the average increase in power [W] due to the production of ice by subtracting the average power drawn without making ice (from step (1)) from the average power drawn while making ice from step (2).

6) Calculate the average energy [Wh/batch or Wh/mass] attributed to the icemaker operation by diving the average increase in power found in step (5) by the average ice production rate found in step (4).

This method of analysis yields repeatable and reproducible results. The data sets required for the ice making portions of this rating method will have to be longer in duration than those required for the steady state non-ice making test already in place. It is difficult to state an absolute amount of time that would be applicable to all units because this aspect of the test is dependent on the influence that ice making has on the overall energy use and is therefore specific to each design. A 24-hour test, as proposed by AHAM (2009) would be more than sufficient, as this amount of time was seen to be adequate for the most extreme case used in this study which had an unusually low ice production rate. The energy use of most units can be quantified very accurately with data that only encompasses 6 or 7 ice making cycles. However, it is important to note that some units may not be able to operate continuously for 7 ice making cycles due to periodic defrost events and full ice bin conditions.

6.2 Other Aspects of the Icemaker Energy Test Method

One remaining issue for consideration is that of interpolation. In this study, we examined each unit operating under several thermostatic settings and analyzed the data with and without interpolating the results of the different measurements. The goal was to determine whether a more useful test result would be produced by interpolating between the measurements of multiple tests. The results showed that the ice making energy for each unit did not change significantly at different thermostatic settings as long as the temperatures in the compartments did not change significantly in response to the icemaker operation. If the temperature in one compartment increased when the icemaker began producing new ice, a single point test could produce an unrealistically small value for the icemaker energy use.

This temperature response is common for units that employ simple, mechanical thermostats. For these units there are benefits and drawbacks for comparing the results of measurements that yield significantly different compartment temperatures. On one hand, the resultant temperature change would actually occur during field use if the unit were to suddenly and temporarily operate the icemaker. On the other hand, there is no standardized point to which the test results from different units could be compared; and there is no limit to the extent that a unit could alter its internal temperature which could potentially result in zero or even negative energy use for the icemaker. Furthermore, if such a unit were used to produce large quantities of ice during field use, it is likely that the temperatures would be adjusted by the user to ensure safe operation of the product. This is similar in nature to the rationale for using the standardized target temperatures of 3.9 °C (39 °F) in the refrigerator compartment and -17.8 °C (0 °F) in the freezer compartment; i.e. these are not by any means the required settings but rather an expectation of the temperatures that users would specify during normal field operation.

There are essentially three different options for addressing this point. The first option is to use a single point test and simply accept the credit that would be given to units that operate in this manner, since this type of operation may actually occur during field use. The second option is to take additional data at a different thermostatic setting and interpolate the results. This option would include a larger test burden but would provide a result that is more adequate for comparing the operation of different products at the same rating condition. Finally, the third option is to use a single point but require the temperature deviation to be smaller than a predetermined amount; i.e. the test operator may have to change the thermostat settings so that the temperatures measured during the test would be within a specified range of the temperatures measured during the non-ice making portion of the test.

Furthermore, if a single point test is selected, it is important to consider the implications for a unit with an electronic thermostat. Units with electronic thermostats typically operate in such a manner that they do not realize a significant change in temperature when the icemaker is operating, but they could be easily reprogrammed in order to capitalize on the credit that a mechanically controlled unit would receive. Without very carefully structuring the text defining a single point test method, a unit with an electronic thermostat could exploit the test procedure to receive an unrealistically low value of icemaker energy use.

7: References

AHAM, 1979. ANSI/AHAM HRF-1-1979, American National Standard for Household Refrigerators and Household Freezers. Chicago: Association of Home Appliance Manufacturers.

AHAM, 2009. AHAM Update to DOE on Status of Icemaker Energy Test Procedure. November 19, 2009.

ASNZ, 2007. ASNZ 4474.1 Performance of household electrical appliances— Refrigerating appliances. Part 1 Energy Consumption and Performance.

Uniform Test Method for Measuring the Energy Consumption of Electric Refrigerators and Electric Refrigerator-Freezers, 10 Federal Register 430, Subpart B, Appendix A1 (01JAN2010), pp. 159-167.

Haider, I., Feng, H., and Radermacher, R., 1996. Experimental Results of a Household Automatic Icemaker in a Refrigerator/Freezer. ASHRAE Transactions: Symposia, SA-96-7-3, pp. 541-545.

Meier, A. and Martinez, M., 1996. Energy Use of Ice Making in Domestic Refrigerators. ASHRAE Transactions, Vol., 102, Pr. 2, pp. 1071-1076.

Yashar, D. and Park, K., 2011. Energy Consumption of Automatic Ice Makers Installed in Domestic Refrigerators, Revision 1," NIST Technical Note 1697, NIST Gaithersburg, MD USA.